TIME, SPACE, AND SOCIETY:
GEOGRAPHICAL SOCIETAL PERSPECTIVES

T0332242

The titles published in this series are listed at the end of this volume.

Time, Space, and Society: Geographical Societal Perspectives

by

Aharon Kellerman
*Department of Geography,
University of Haifa, Haifa, Israel*

KLUWER ACADEMIC PUBLISHERS
DORDRECHT / BOSTON / LONDON

Library of Congress Cataloging in Publication Data

Kellerman, Aharon.
    Time, space, and society.

    (The GeoJournal library)
    Bibliography: p.
    Includes index.
    1. Anthropo-geography--Philosophy. 2. Space and
time. I. Title. II. Series.
GF21.K45 1989   304.2   88-36399

ISBN 0-7923-0123-4

Published by Kluwer Academic Publishers,
P.O. Box 17, 3300 AA Dordrecht, The Netherlands.

Kluwer Academic Publishers incorporates
the publishing programmes of
D. Reidel, Martinus Nijhoff, Dr W. Junk and MTP Press.

Sold and distributed in the U.S.A. and Canada
by Kluwer Academic Publishers,
101 Philip Drive, Norwell, MA 02061, U.S.A.

In all other countries, sold and distributed
by Kluwer Academic Publishers Group,
P.O. Box 322, 3300 AH Dordrecht, The Netherlands.

Printed in The Netherlands

FOR MY DEAR PARENTS.

CONTENTS

LIST OF TABLES

LIST OF FIGURES

PREFACE

Time and space are two of the most basic dimensions of human life. They envelop all human beings from birth to death. As such, they provide the context for human existence. At the same time, however, time and space also serve as major influencing factors in mankind's actions. Hence, a vast literature has developed on time and space as separate dimensions, and recently on time-space as joint dimensions. Interestingly enough, the social connotations of time and space have mostly been studied with the individual human being in mind. The more societal significance of time and space, whether separately or jointly, have been relatively neglected. It is the purpose of this volume to help fill this lacuna through discussions on some of the many junctions of time, space, and society at large.

The discussion will naturally involve concepts and findings from more than just one discipline -- notably, geography, sociology, social history and political science. It is, thus, obvious that the topic may be highlighted from several perspectives. Given my own education and work, the approach will lean more to the geographical perspective. Geography has a special merit as an integrating framework for the study of time, space, and society. It is a discipline that has space at the center of its raison d'etre and, as such, has always striven for integration, holism and comprehensiveness. Geography, therefore, has absorbed concepts and theories from other social sciences in symbiotic and sometimes eclectic fashion. The unique importance of the geographical viewpoint has not always been appreciated. As the denoted historian Fernand Braudel put it: "Spatial models are the charts upon which social reality is projected, and through which it may become at least partially clear; they are truly models for all the different movements of time (and especially for the longue durée), and for all the categories of social life. But, amazingly, social science chooses to ignore them."[1]

The theoretical framework for the study of the changing concepts and uses of time and space originated in sociology rather than in geography. The newly developing theory of structuration permits the study of the ongoing dialectics between society at large, on the one hand, and time and space, on the other. It also enables comparative analyses of time and space. The theory of structuration, which will be briefly described in the next chapter, is based, among other elements, on geographical notions of time and space. It has already been applied to the geographical study of both individuals and society at large.[2]

The chapters of this book will be presented in the following order. First, the idea of societal time and societal space will be elaborated, mainly through comparisons to the time and space of individuals. Next, societal time will be compared to societal space at the theoretical level. The possibility of a homology between the two will be explored, and common terminology proposed. Third, societal time and space will be compared at the urban level, with a focus on the changes in cities in capitalist systems since the industrial revolution. The analysis then turns to the two most basic social groups, men and women. Evidence for different conceptions, perceptions, and uses of time and space will be presented within the framework of modern society, and the possible implications for urban life will be explored. Finally, the focus moves to a national arena, the transitions in spatiality and temporality within Israeli society. The choice of Israel is not only linked to my own nationality, for Israel is also an excellent setting for a national analysis of time and space because of the many social and geographical changes that this society has experienced in the twentieth century.

My interest in the systematic study of time-space rose in the mid-1970s, when I became interested in the linkages among the spatial expressions of growth in the urban fringe, the time it takes for them to develop, and the growth processes that operate in both time and space. This interest developed during my Ph.D. studies at Boston University. Later, at the University of Miami, when I became exposed to Hagerstrand's time-geography, I tried to see whether his conceptual framework could be applied to the urban and regional levels as well.[3] Using a model developed by Krakover, we also tested this approach in part on some data for Philadelphia.[4] I first examined the idea of a time-space comparison at the societal level in 1982-83, during a sabbatical leave, from the University of Haifa, at the University of Maryland, College Park. It was then that I discovered Giddens' structuration theory, which led to my work on time-space homology and on time versus space at the urban level.[5]

After presenting an analysis of time versus space at an annual meeting of the Israeli Sociological Association, I was confronted with criticism that the analysis applied mainly to men and not to women. This brought me to a study of time-space differences between the sexes and their possible implications. Simultaneously, I became involved in a study of transitions in spatial patterns and social processes within twentieth century Israeli society. One of the questions that begged consideration had to do with the changes in temporality and spatiality that occurred in this society -- and led to the writing of the fifth chapter in this book.

Excited as one can be about an evolving research topic, it always involves help from others, and this study is no exception. Research and writing took place at the University of Haifa and at the University of Maryland. I wish to thank my students in Israel and in the U.S. whose comments were most beneficial in many ways.

I owe special thanks to Kenneth E. Corey, Chairman of the Department of Geography at the University of Maryland, College Park, for providing the necessary facilities and atmosphere for me, while serving as Visiting Lecturer there, several times between 1982 and 1987. I received also generous support in this regard from the Research Authority and the Faculty of Social Sciences at the University of Haifa. David Gross, Department of History, University of Colorado, provided some very useful insights on the full manuscript, while separate chapters were read by Robert D. Mitchell, Nurit Kliot, and Orna Blumen. I used many of their suggestions, but responsibility rests, obviously, with me.

At a different level I wish to acknowledge permission granted by Tijdschrift voor Economische en Sociale Geografie, for the use of my papers, which were published there in 1981 (for a section in Chapter 1), and in 1987 (for Chapter 2).

The typing of separate chapters and of the full manuscript was shared among some most devoted and efficient typists: Pat Leedham at Maryland, Danielle Friedlander, Heather Kernoff, and Genoveba Breitstein at Haifa. I owe them much gratitude for their efforts. Copy-editing was professionally and carefully performed by M.A. Goldstein. Artwork was skillfully designed and executed by Aliza Gold at Haifa.

Last but not least, I received both active and passive support from my wife Michal, and from our daughters, Tovy, Miriam and Noga, whose patience, understanding and sharing made it all possible.

NOTES TO PREFACE

1. Braudel, 1980, p. 52.
2. E.g. Moos and Dear, 1986; Dear and Moos, 1986.
3. Kellerman, 1981.
4. Kellerman and Krakover, 1986.
5. Kellerman, 1987b.

CHAPTER 1

INDIVIDUAL AND SOCIETAL TIME AND SPACE

"Time flows; yet it is always present. That is why the problem of Time has tormented the minds of all great philosophers far more than its correlative, Space."[1] Geographers would reverse this order of attention, although they might admit that geography deals with the man-environment system principally from the point of view of space in time.[2] This approach to time, however, was evolving, so that by the mid-1970s a geographer could claim that "the essential unit of geography is not spatial - it lies in regions of time-space and in the relation of such units to the larger spatio-temporal <u>configurations</u>. Geography is the study of these configurations."[3] This last identification of the geographer's realm of study can be recognized as a result of the impact of Hagerstrand's time-geography, and, to a lesser degree, of Chapin's activity-patterns approach, both of which have evolved during the last twenty years.[4] Their work has led another geographer to declare that time-geography "represents the most significant thing happening along the frontiers of human geography."[5]

Hagerstrand's approach emphasizes the contextual time-space framework of human actions through treating both dimensions as limited resources, while Chapin's more empirical inductive approach refers to time as a factor in human actions.[6] These two methodologies and their derivatives have in common a view of time as a dimension related to the individual and to groups of people. Although such an outlook helped to bring geography "back to the people," it had at first relatively little direct impact on main-stream human geography, especially urban-economic geography, which had at its center the task of studying the spatial allocation of man-made activities rather than of studying human beings themselves. In this latter, more "traditional" field of study relating to the structure of city and region, time has not been completely ignored, especially since geographers have started to look for "dynamics" and "processes" of spatial order. Only a handful of approaches, however, have been developed in order to understand the significance of time in spatial structure and the changing significance of space in time, and no integrated time-space approach has been suggested for the many processes found to be at work in cities and regions.[7]

Time and space may be viewed as being both contexts and compositions. As the most basic dimensions of human life, they serve as obvious, almost trivial, contexts of human life. On the other hand, time and space are also compositional in their being resources and factors for human action. This dual nature of time and space lies at the basis of the discussions in following chapters. It is thus imperative to examine at the outset the use of contextual and compositional scientific "languages" for the study of time and space.

Time and space: context and composition

Differentiating between a contextual and a compositional study draws on a slightly more generalized debate concerning research differentiation: space-time language versus substance language.[8] In the first language, objects or events are studied by reference to their four-dimensional "contextual" existence (x,y,t,z, standing for longitude, latitude, time, and depth, respectively, of an event, a point, a line, an area or a volume); in the second language, they are studied by their properties relating to "compositional" structure and form $(p_1,p_2,p_3,...,p_n)$. The latter language, highly comprehensive in terms of what and how to study, is used by most scientists. The former, however, may be used in three forms:

> The first [form] conceives of the individual as constituting a space-time region (i.e. an individual is defined by the spatial extent of the thing over all time periods). The second conceives of the individual as forming a space region without any temporal extent (regionalisation at one point in time is a good example of this). The third regards the individual as a series of space-time points (this provides a way of handling phenomena varying continuously over space and time - a climatic region might be built up in this particular language). These three forms of language can be developed and expanded, but they have not been given a great deal of attention in the philosophical literature. To the geographer an understanding of their properties and relationships is vital, for, whether we like it or not, we are forced to make use of such languages.[9]

It is obvious that a contextual time-space approach uses the third form of time-space language, which enables a joint treatment of time and space.

The Geographic Matrix proposed by Berry was an attempt to unify compositional and space-time languages in geographic research.[10] The matrix consists of three dimensions, namely of places, characteristics and times. Each cell would, thus, contain a geographic fact, or a characteristic (composition) at a given place and at a certain time (context). The matrix may be sliced along times, places or characteristics. Geographers who mixed context and composition were, however, criticized for their use of "two rather different languages possessing rather different characteristics and it may well be that many of our methodological problems arise from failure to understand the properties of such languages."[11] Even within the use of contextual language in geography too much emphasis was put on space. Hagerstrand, in introducing his time-geography, claimed that in "the traditional geographical format", there was a "very strong emphasis laid upon the thin spatial cross-sectional view of the flow of terrestrial events."[12] The charge received support, including the suggestion that "it is a challenge to cease taking distance itself so seriously."[13] In other words, geographers had been overemphasizing the second form of the space-time contextual language, while virtually ignoring the temporal dimension.

The study of time-space must, however, use the languages of both context and composition at the same time for several reasons. First, space-time (context) cannot be separated from its substance (composition). The space-time study of any spatial organization will remain incomplete if not coupled with an explanation based on the nature of the object or the event and expressed in substance language. Second, space and time can only be studied together by using the third form of space-time language. This contextual study of process in space has to be pursued alongside an understanding of the structure and meaning of time and space by their users. Third, each process studied has to be viewed as being part of a longer chain-process. These processes involve both contextual changes of time-space and compositional transitions in their uses and social significances.

Time and space may thus be classified most basically into absolute and relative time and space. In absolute time and space, the two dimensions are viewed as passive contexts or as frameworks and containers for human life. Relative time and space refer to the active use of these dimensions and resources by individuals and societies. They further relate to the reshaping by individuals and societies of time and space as containers.

Time and space studies may be ranked along with their classification, as active and passive. This may be done using a linguistic ranking of primitive concepts and syntactic, semantic, and pragmatic levels[14] (Figure 1). The primitive notion of space consists syntactically of two levels: absolute and relative space. Relative space leads to locational analysis in both its contextual (form two of space-time language) and compositional-explanatory aspects, resulting in theories and hypotheses concerning distance. These have become the focus of study in positivist human geography.[15] Our concern here will be, rather, the spatial aspects handled by contemporary social theory: namely, the changing significance and uses of space by societies and by individuals. Absolute space, which is the perspective used in traditional regional geography, assumes no reference to theories on the structure and form of "ideal" or integrated regions.[16] Similarly, in absolute time, no theorizing is possible about an "ideal future," and no theory exists as yet of an "ideal past." History, therefore, is a non-predictive study of time, equivalent to a regional geography of space. A relative perspective of time permits the formulation of theories on the time needed for given spatial changes to take place, and the obverse: the spatial changes that may occur within a given time. A social theory perspective on relative time would inquire into the changing significance and uses of time by individuals and societies. This inquiry constitutes the focus of the chapters that follow.

A lively, more extensive discussion of time-space studies at the macro-scale comes, through a refined and imported time-geography, from sociology. The theory of structuration, in the version proposed by Giddens, employs time-space as one of its major bases.[17] Structuration theory argues, simply, that societal time and space are indispensible elements for the study of the constantly ongoing interrelationships of human agency and social structures. The several aspects of time and space that will be highlighted immediately below and in the following chapters will be treated within a structurationist framework. It is, thus, important to outline at this point some basic relevant notions of

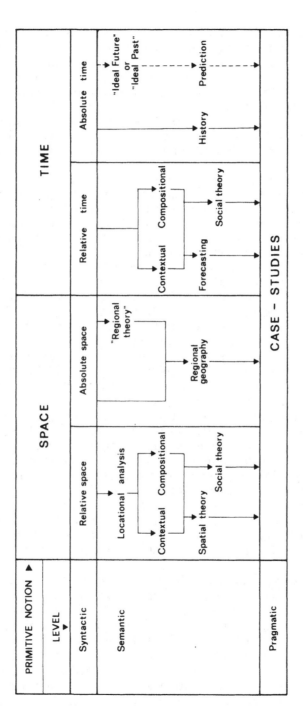

Figure 1:   Levels of Theoretical Studies of Time and Space

structuration (though some other aspects as well as criticisms of structuration will be left for later chapters).

Structuration and Geography

The theory of structuration was named as such by Giddens, though its development has been shared in by others, as well.[18] Essentially structuration attempts to bridge between structure and action by proposing a "duality of structure" in which "the structured properties of social systems are simultaneously the medium and [the] outcome of social acts." The concept implies that structures are constantly undergoing changes. Both social structure and human agency may be studied through their location in time and space, which, in turn, undergo continuous transition in their social uses. Structuration is a general approach of holistic nature; as such, it incorporates structure and action, intention and practice, context and composition, individuals and societies, and constraints and enabling factors.[19]

In several of its notions and implications, structuration presents a challenge to geographers. First, it is a social theory that explicitly admits to the crucial social role of time and space. Second, the attempt to connect structure and process should appeal to geographers, who have had a long tradition of attempts at holism. Third, structuration may prove useful for an integration of geographical patterns and social processes. Fourth, structuration could provide an integrating framework for humanistic and structuralist approaches in geography.[20]

Explicit references to structuration in the geographical literature have become frequent.[21] The review by Thrift could be interpreted as proposing an "export-import" relationship between geography and structuration. On the one hand, geography "exported" several notions, especially time-geography, to structuration, while the latter could, and perhaps should, serve as a framework for a new regional theory in geography. The geographical scale is, thus, regional, and the new theory should consist of elements pertaining to individuals (e.g., personality) as well as societies (e.g., ideology). By contrast, Pred emphasizing the place scale and the individual, carried the path and project notions from time-geography into structurationist-geographical analysis.[22] Gregory put a third dimension -- society -- at the center of his structurationist analysis of industrial change within a regional framework.[23] His study of structuration was not just theoretical, but presented an empirical application to geographical analysis, as well.

Since the crux of structuration is the constantly evolving relationships between individuals and society, its application to geographical study may take, in any given study, at least one of three forms. It might, following Pred, emphasize the roles of individuals in social and geographical change: it might focus on societies and geographical change, as in Gregory's study; or it might center on the regional milieu, as proposed by Thrift. Potentially, too, there may be an attempt to study two or all three of the dimensions simultaneously. The focusing on only one dimension within a theoretical framework that attempts to bridge among the three was sometimes seen as disturbing when it came to empirical study.[24] Giddens termed the focusing on one dimension "bracketing", and declared society at large, on which this volume deliberately focuses, to be the most important dimension for social theory.[25]

Gregory defined geographical structuration by stating that "man is obliged to appropriate his material universe in order to survive and [because] he is himself changed through changing the world around him in a continual and reciprocal process."[26] Harvey moved this definition one step forward: "Geographical knowledge records, analyzes and stores information about the spatial distribution and organization of those conditions (both naturally occurring and humanly created) that provide the material basis for the reproduction of social life. At the same time it promotes conscious awareness of how such conditions are subject to continuous transformation through human action."[27]

This definition is interesting when interpreted in the light of formerly widely accepted definitions of human geography. Generally speaking, until the 1970s, the major concern of human geography was space, and Man's impacts on it. Man was mainly a macro or meso-entity, such as rural or urban communities and nations. Space was similarly interpreted as respective geographical units, such as villages or cities. In the seventies, the scope of both the definition and its components were widened. Human geography was related to as the underlineinterrelationships of Man and space. Man could now be individual human beings as well as institutes or structures, while space consisted of micro-units, such as homes, as well as the more macro communities. Still, Man's impact on space was more central than were spatial impacts on society. Harvey's definition reverses this order by according first consideration to the spatial impacts on social life. In addition, Man is represented by "social life" in the first half of his definition, and by means of "human action" in the second half. Space is repalced with "the spatial distribution" of conditions for social life. The constant and dual relationship between humans and space proposed by Harvey for geography is structurationist although it ignores the temporal dimension. The interrelationships of society, space, and time simultaneously determine and are influenced by, the social importance of time and space. And indeed, these constant interrelationships inform our own discussions of time, space and society.

The preceding sections have attempted to lay down several bases for joint discussions of time, space, and society: first, the necessity to link time and space, especially in geography; second, the possible treatment of time and space as both contexts and compositions, relative and absolute; third, the two-way relatinoship between time-space and society. We now move to comparisons between individual and societal times and spaces. These comparisons will provide a rationale for the rather societal focus of this volume and will permit some further conclusions about the nature of individual and (or versus) societal times and spaces.

Individual and Societal Time

Time may mean many things for individuals. It might be an experience, a major dimension, an ordering framework, an event of biological significance.[28] Experiential time, or lived time, refers to personalized images of time as being short or long, passing fast or slowly. Time is a major dimension along which all events occur and around which human lifecycles evolve. It is an ordering framework for events in terms of "before" and "after" and in terms of chains of events or developments. The calendar and the clock are two basic tools in the organization of daily, weekly, monthly, and annual flows of events. Biologically we all have our biorhythms, which regulate organisms and activities. There are at least two forms of inner perceptions of time: (1) as an uninterrupted duration; (2) as a series of differentiated occurrences along past, present, and future. These two internal perceptions of time will be elaborated upon in a later chapter, since they permit the indication of a difference between men and women in regard to time.

For both individuals and society time may be considered as being both passive and active. This duality has been described in these terms: "There is an obviousness to the notion of time timing. it clocks, it charts, it historifies, it measures, it numbers. As such, time is viewed in a passive role or better yet, as a silent agent, as an effect. It does what it is told. Combined, however, with this aspect of time, there is the aspect of time as a cause. Time is not relegated to silence, to passivity. It causes, it directs, it generates."[29]

The notion of "time of society" has not been treated and developed in a systematic manner either by geography or by sociology. Although "social time" has been the focus of several sociological works employing different perspectives, the approaches used lacked a clear, detailed distinction between the time of individuals and the time of societies.[30] One could argue, at least from a sociological viewpoint, that these two times represent a continuum rather than a dichotomy and that the two are highly interrelated. The argument has been phrased this way: "perhaps, following Sartre, we should attempt to comprehend 'present' circumstances both progressively and regressively, by going back and forth from social totality to the individual and from the individual to the society in search of the groups (such as the family, school, association, church, class and state) which mediate between the individual and society."[31]

Though some studies differentiate between individuals' time and society's time, they do so only for definitional purposes rather than for contrasting the two or at least distinguishing between the two in detailed examples. Gurvitch, for example, talked about "micro-sociology" and "macro-sociology," but when it comes to his analysis of the "global-society," it is not clear which sociology he attempts to address. By the same token, Fraser defined the "sociotemporal Umwelt" as the time of global collective Man, compared to five other kinds of times that are those of the individual; but his detailed analysis concentrates only on individuals. Durkheim, too, distinguished between private time and time in general, an approach that led to a distinction between private and public time.[32] On a more abstract level, Luhmann defined his "world time" as "an infinite series of temporal points compatible with the assignment of different values to particular points in different system histories.... It is a dimension of the horizon of the world. Because it is measured uniformly, world time allows processes in <u>all systems to run simultaneously.</u>"[33]

It is notable that all the terms mentioned above provide for <u>dichotomies</u> between individual time and societal time. Interesting in this regard is Bergson's attempt to uncover <u>processes</u> by which societal times are internalized in individual times.[34] The human mind tends to use understanding rather than feeling, thus distorting the true <u>duree</u> into disparate entities. True <u>duree</u> though, may be reached through intuition. (We shall return to this important distinction in Chapter 4, when discussing differences in time perception between men and women.)

Giddens' writings on social time are interwoven in his discussion on the nature of society and on differences between it and individuals. Thus he rejects the view of society as a "functional unity of parts" although he uses the terms "society" and "collectivities" interchangeably. "Micro" and "macro" sociologies are different, according to Giddens, in that the latter represents "interaction with others who are physically absent (and often temporally absent also).[35] "Society" is thus, a nation-state, characterized by a locale (social space), an "institutional clustering of practices" and "an over-all awareness -- of belonging."[36]

Among economists, the study of societal time as a social resource has not been extensive. Two reasons for this lack of emphasis have been noted. The first is the economists' recognition of production factors only when in scarce supply; the second is the long associaton of time with labor supply.[37] At the macro-level, Soule stated the need to study time as a scarce resource in coordination with land and other production factors, but he stopped short of providing such an analysis. Some micro-econometric models relating to the consumption of time at the household level have also been developed.[38]

Using here a structurationist framework, we shall relate to time of society along Giddens' views. In order to differentiate it from individuals' social behavior in time, the term "societal time" will be used for time of society. (The many possible relationships between the two are not meant to be excluded, but are beyond the scope of this chapter.) Furthermore, society will be dealt with here in both its national and urban/regional contexts, with an accent on the latter. These two spatial contexts serve as the arenas for most economic and social activities in capitalist systems.

The crucial importance of scale distinction when dealing with time-space was suggested by Parkes and Thrift.[39]

> Linkages in space-time that can be readily identified, and possibly explained, at one scale, remain abstruse, even irrelevant, at another scale. It is also necessary to establish the right correspondence of scales; in some cases the spatial scale may be right and the temporal scale incongruous, and in other instances the inverse may be the case or both may be wrong!

The problem in applying such a pronouncement to more detailed research (and Parkes and Thrift did not attempt this) lies in the assertion that "in the social sciences definition of space and time is non-existent."[40] Nevertheless partial definitions do exist, as in Shackle's discussion on economic relations and Lynch's notions concerning architectural environments.[41] If, however, we follow the distinction made here between the study of human behavior in time-space (the human scale), on the one hand, and the study of societal time-space (the societal scale), on the other, then it is possible to identify some basic differences and similarities between the two scales. The comparison that follows is not supposed to be exhaustive, but follows the characteristics of time as suggested in the literature devoted to human behavior. The discussion starts with the differences between the two scales and proceeds to the similarities.

One of the cornerstones of time-space frameworks is the assumption that time is a limited resource. Time is finite in terms of its availability to a human being; therefore, it should be dealt with economically in the same manner as all other production factors.[42] It is difficult to apply this notion to the study of society, since it would seem illogical to assume that society has a limited time-span to develop and change in space. It is true that certain development plans sometimes call for spatial change to take place within limited time boundaries; five-year plans are an example. But it is also true that lagging behind schedule is possible from the viewpoint of time availability for society, whereas the time an individual might use is limited by his or her life span. The time limits of individuals may sometimes be aggregated into a population time limit, as in certain travel studies, like one that suggested that "during a fixed interval of time, a population of given size has a particular amount of time at its disposal."[43] In the long run, however, society acts in space on a "generation after generation" or an unlimited time-basis, while individuals act in a limited time-space environment.

It also seems that time is often used by individuals to adapt to a given spatial organization. This is one of the bases of Chapin's action spaces and Pred's time-space framework for innovations.[44] Society, on the contrary, uses time in order to alter a given spatial organization of its activities. This is performed either through or without formal institutions. It may be argued that these two processes are mutually complementary, that is to say, that a long-run process of adaptation of individuals to a given spatial organization creates changes in that organization, and vice versa. It is still important, however, to realize the difference in attitude toward time in the space of individuals, in their daily life, and in the long-run development of society. Thus, society has been shown to have, through long-range planning, a broader future horizon than does the individual as well as longer past experience.[45]

Another notion developed for the study of human time-space behavior is the idea of the "colonization of time."[46] Melbin shows that modern urban society invades and colonizes the night hours in the same manner as Americans moved westward in the last century. This idea has its roots in the conception that time is a usable and finite resource although Melbin claims that "space and time together form the container of life activity."[47] This interesting analogy is only partially relevant at the societal scale, since changes in the spatial organization of society evolve on an eternal calendar rather than on a finite, 24-hour-clock basis. Thus time is always used, though there are slower and faster periods of change of any kind. Society may try, however, to make more efficient use of time in space by using 24 hours a day, a point to which we shall return in a later chapter.

Another form of colonization of time that pertains to both individuals and societies is the longer life expectancy in modern societies. This makes for the need to plan for a longer "third age" by both individuals and societies. It explains, for example, some of the growth of the U.S. sunbelt and the evolution of city-sections with concentrations of older people. A third characteristic of time for the individual lies in its being a personal experience; in Rose's descriptive trio, time may be "lived," "subjective," and an "ingredient."[48] Rose argues that Hagerstrand's view of time is a framework rather than a phenomenological-experiential idea. Again it is obvious that society does not "feel" time as a psychological experience although certain elements of time-experience may be related to macro-scale time, too, when an inter-society comparison of time-use is being made, based on differences in social values and constraints. The same is true when differences between men and women regarding the perception of time and space are traced.

Parkes differentiated among three types of time: biological, psychological and socio-ecological. These were related by Thrift to four levels in the social system: superstructure, built environment, activity systems, and attitudes and perceptions.[49] Spatial order (in which the "built environment" is included), then, is not <u>directly</u> influenced by biological rhythms and psychological motives and constraints. The weight of socio-ecological time, on the other hand, is much larger in societal time-space than in an individual's time-space.

In addition to the differences between time at the societal and human scales, several similarities can also be identified. In both scales, time is more than just "a ladder for processes to climb on" and more than "a locational and co-locational continuum."[50] It is foremost a basic resource, and at the societal scale it is also a major factor determining spatial change. This factor, or in other words, the time needed for some spatial change to happen, differs from one system to another (e.g., the time needed for the suburbanization of commerce would not necessarily be the same as the time needed for the suburbanization of manufacturing.[51] Moreover, the time needed for any spatial change in a given system in one society is different from the time needed for the same change in another society (e.g., the suburbanization of commerce in North America <u>versus</u> Europe). The spatial reaction of an economic system in one country to an economic impulse may be but an instant if the intra and/or intersystem transmission of impulses is fast; whereas in another society these processes may be slow and, therefore, spatial adjustment to certain economic changes will be slow. By the same token, private entrepreneurship in one society may be encouraged, and many people will therefore take a chance on using risky locations in view of some evolving or expected economic change; in another society, taking economic risks may not be a social goal (or the government might play a larger role in the space economy, thus reducing risks).

The potential availability of several social modes of spatial-economic behavior is similar to the approach adopted by students of human spatial behavior; namely, that individuals use time in a manner reflecting their personalities. This point has been related to as follows:[52]

> Any culture at any period in its history appears to be characterized by a particular space-time coefficient. Time budgets will reflect cultural objectives, technologies, resources; the geography of the intensity of occupance by residence times will be a mirror image of the latter. Any change in, say, the friction of distance will be reflected in time budgets.

As a result, the term "relative time" has been suggested for the different degrees of time-use by several societies:[53]

> With the important basic constraints recognized, it is then necessary to study how different societies and groups embroider the basic time-space framework, through specific rules, cultural traditions and so on, to their advantage (or otherwise, if they have no control over such things). 'Absolute' time in this framework is a mixture of physical limits to action and agreed societal referents, whereas 'relative' time essentially arises out of that part of the temporal structure which is open to change.

Thus, we may identify "tempral intensity" as being comparable to spatial intensity concerning the use of time by society. If "relative" time is large, then attempts to intensify time-use will be low, and vice versa.[54]

The time-space approach to human spatial behavior maintains that time-space use is limited to those constraints defined by Hagerstrand as capability, coupling, and authority constraints.[55] Capability constraints relate, for example, to the limited reach of the human body, or to the need to devote time for sleeping. Coupling constraints refer to the cooperation and coordination between two or more people in order to perform certain tasks. Authority constraints are, for example, "no entrance" road signs that limit spatial (and temporal) movements. At the societal level, the "use" of time by a certain society for some spatial change or for the adjustment of one spatial system to a change in another is mainly limited by authority constraints. These may take many forms: There are master plans, zoning laws, etc., at the practical level; the eocnomic-political system at the legal level; and national (or religious) ethnological characteristics at the cultural level.

In another study, Hagerstrand recognized eight basic limitations of human time-space behavior. They may be defined as follows:[56]

1. the indivisibility of the human being (and of many other entities, living and non-living);
2. the limtied length of each human life (and of many other entities, living and non-living);

3. the limited ability of the human being (and of many other indivisible entities) to take part in more than one task at a time;
4. the fact that every task (or activity) has a duration;
5. the fact that movement between points in space consumes time;
6. the limited packing capacity of space;
7. the limited outer size of terrestrial space (whether we look at a farm, a city, a country, or the Earth as a whole); and
8. the fact that every situation is inevitably rooted in past situations.

Of these limitations, the first four have no meaning at the societal scale, but the last four do. For societies, too, there is a limit to space as a resource, there is a past time behind every present, and movement consumes time. Carlstein mentioned the notion of "positive constraints" in general, but the stimuli for both human individual and societal action provided by time and space have still to be explroed. But neither does Giddens, who voiced this criticism, put forth any positive thesis about the source of human action.[57] The basically constraining view of time-space by the Swedish school has been attributed to the more restricting connotation of the Swedish word, rum (space), compared to the openness and boundlessness of its English equivalent -- space.[58]

At both scales, some importance attaches to a chain effect; i.e., the interdependence of events. The movement of an individual from one place to another may cause a second mvoement, which might be of a cyclical nature (commuting, for example) or of a cumulative nature (residential change). At the societal scale, any spatial event may be interpreted as being dependent on a former event; thus we may mostly refer to change as an accumulation of some smaller changes or as a link in a longer chain (both temporal and spatial). These societal, macro-scale interdependencies are, therefore, mostly cumulative rather than cyclical.

A major hidden assumption of the Hagerstrand approach to individuals' time and space is that the two are movement resources, restricted in nature. As such, the conception of time and space is contextual rather than compositional. This might be the reason that people's activities are assessed independently of their social settings, and why there is no consideration of power.[59] For society at large, however, time and space are more than merely movement resources. Above all, they are production resources, major organizational dimensions, and containers. If time and space are assumed to be the major movement constraints of individuals, then time and space are not only homologous in nature, but inseparable as well. The question is whether this may also be said about the nature of societal time-space, which has production and organizational dimensions, too. Moreover, the additional significance of societal time and space may entail different relations between them from those between individuals. These questions will be discussed in the next chapter, which is devoted to the time-space homology.

Another area of similarity between the human and societal scales is in the nature of time. Following Rose's comments on Hagerstrand's time-geography, it is possible to recognize time at both scales as being abstract, a priori defined, and mostly non-experiential. In addition, time at both scales follows a Western conceptualization so that it is orderly, linear, uniform, unidirectional, irreversible, and homogeneous.[60]

Individual and Societal Space

As with time, one may differentiate between individual space and societal space, and compare the two. The study of individual space developed within the subdisciplines of behavioral geography and environmental psychology, especially since the early 1970s. It usually has not concentrated on the study of small spaces, such as rooms and homes per se, but rather on their interrelationships with their users, mainly individuals. Thus, studies have concentrated on such aspects as movement about small spaces, their organization and use, their image, and constraints. Some of the major terms are outlined below.

A differentiation may be made between two forms of individual spaces. On the one hand, there is the more fixed territoriality, the attachment of people to spaces such as rooms and homes; on the other hand, there is the rather more dynamic personal space, which constantly surrounds every human being like a bubble.[61] We may, thus, identify three levels of individual spaces and their patterns of behavior: microspace is equivalent to the "bubble like" personal space; mesospace usually relates to the home or neighborhood; and macrospace is the home range or the area of commuting, shopping, social interaction, etc., around the home.[62] Macrospace actually consists of two kinds of space. One type is action space, defined as "the collection of all urban locations about which the individual has information, and the subjective utility or preference he associates with these locations."[63] Within this action space lies a second kind, the individual's activity space, or the spatial area within which actual activity takes place (i.e,. macrospace).[64] Action space is determined by many factors, among which are an individual's values. These, on their part, have to do with social interaction in space, or social space.[65]

Interestingly enough, the "space of society", or societal space, has not received thorough treatment or wide terminology by geographers as has been the case with individuals' space. Although large, human-shaped spaces, such as cities, regions and countries, have traditionally been the domain of inquiry for human geography, the emphasis has usually been on the spatial organization of human-made artifacts, such as homes, roads, factories, etc., and less on the interrelationships of society and space. Two exceptions should be mentioned in this regard: the more traditional study of the creation of a national identity of territoriality and the more modern study of the sociospatial dialectic. The first of the two subjects refers to the territorial aspect of statehood. Thus, Hartshorn identified centrifugal (or disturbing) and centripetal, (or assisting) forces at work in the creation of a national identity of territoriality. Gottmann emphasized the role of accessibility in the shaping of societal territorial orientation, and Knight discussed territorial attachment at the level of nation and region.[66] These theories may explain economic development and political attachment, but they are not theories of action or intention. Furthermore, they relate to the significance of space as a politcial symbol and tool, thus ignoring social and cultural significances and their interrelationships with social change.

The sociospatial dialectic was proposed by Soja, in order to view society and space within a unified framework, in which the two entities constantly influence each other.[67] Soja's dialectic will be in the center of our discussion on spatiality in the next chapter and will be elaborated there in our attempt to show, within a framework of structuration, the ongoing interrelationships of society and time, on the one hand, and society and space, on the other. In addition, some possible interconnections between societal time and societal space will be identified. Before moving to these discussions, however, further attention should be given to societal space by way of comparison with individual space, in similar fashion to what was done for time. Here, too, the discussion will start with some differences between the two, followed by several similarities.

A major difference between individual time and societal time, discussed in the preceding section, is the finitude of individual time, compared to the relative infinitude of societal time. Duration has, thus, different meanings in these two times. Equivalent "durations in space" would mean the ability to spread activities on chunks of space. This ability is limited for the individual. Although people's activity space consists of home, work place, shopping, and social space, they obviously cannot be in at more than one place at a time. This indivisibility of the human being was mentioned earlier as a capability constraint for action. When it comes to society, whether urban, regional, or national one, this constraint does not apply. Limits, though are set on societal duration in space by authority constraints (boundaries) or by the terrestrial size of space (islands, for example). Still, it may be argued that society exists and performs in more than one point at a time. An urban society, for example, though using a relatively small space, consists of many activities, at any given time, in many small loci of both production and consumption.

Despite the more limited extent of individual space, the individual has an advantage over society, when it comes to spatial dynamics, or transitions in activity spaces, because of human indivisibility. A person's action and activity spaces may shrink, expand or change altogether, quite drastically. These changes may happen because of an increase/decrease in income, professional change, migration, stage in lifecycle, and other events. Transitions in the spatial horizon of a given society are slower and sometimes next to impossible. A nation may expand spatially in two ways, internally and externally. Internally, a new inter-regional equilibrium may be created when some regions enjoy development or when other regions lose population and economic activities. Such processes are normally described in terms of center-periphery relations. Achieving positive regional change is a national goal of most nations, but has become a difficult target, whatever means are used.[68] Externally, a nation may expand into new territories by way of peaceful or war-time annexations. This usually takes place, not in a process-like manner, but in the form of single events; obviously, not all such annexations result in an integration of the new territories with the old ones. Much faster societal spatial expansion became possible for modern urban areas when processes of suburbanization made the city not only larger in population and space, but also provided for a renewed allocation of economic activities and population densities along the extended urban spaces.[69]

A third difference between individual and societal spaces relates to the terms, intensity and density. One may attach the term intensity to both types of space. A farmer may cultivate a field at different levels of intensity, and an urban society may use urban space at different levels of intensity. The term density, however, applies at the societal level only, since by definition it relates to the number of people, or any other unit, on a given piece of land.

In addition to their differences, individual and societal spaces also exhibit similarities. Foremost among these is the fact that space for both individuals and societies is not a mere passive dimension on which all human activities take place. It is a political dimension for municipal and national jurisdictions; it is an economic resource expressed in land prices; it is a constraint for human activity, since it is limited and requires movement efforts and time to reach its several components; and finally it provides for individual and societal experiences.

Space as an experience is an important element. An individual may sense landscapes and places, and create internal images and mental maps of them. These personal experiences relate to natural environments as well as to the aesthetics of man-made urban environments. They have become foci of study within the disciplines of humanistic-cultural geography and architecture.[70] Space also provides, however, for national experiences via national symbolic sites and landscapes. Both personal and societal spatial experiences may be influenced by the territorial size of countries and by their internal regional organization.[71]

For both individuals and societies, space is a limited resource and framework. As previously discussed, this limit might turn out to be less restrictive for individuals than for societies expanding into new territories. Individuals and societies are also restricted by several authority constraints that regulate their movements and uses of space. These limitations may be part of the legal system or may stem from other sources, such as economic-political restraints or cultural-religious values.[72]

Finally, individual and societal spaces should not be looked upon as independent and dichotomous entities, but rather as continuous and interdependent. Individuals' activity spaces are established within societal spaces; on the other hand, new components of individuals' activity spaces create expanded societal ones. By the same token, societal spaces are dependent on individual ones, since they may be viewed as consisting of many, interacting individual spaces. On the other hand, personal activity spaces are dependent on the limits and several components of societal spaces.

Summary and Conclusion

The discussions in this chapter have attempted several objectives that together serve as the basis of the following chapter. The discussion went from the more general to the more specific. First, geographic perspectives on the need for a joint treatment of time and space were briefly presented. Next, the uniqueness of time and space as being simultaneous contexts and compositions was shown. Third, some preliminary notions on the special attitude of structuration theory to time and space were outlined. Fourth, two comparisons were made: between individual and societal time and space, respectively.

Thus, four forms of time and space have been identified; individuals' and societal times and individual and societal spaces. The discussion leaned more toward societal time and space, since they will be in the focus of the chapters that follow. Putting the four forms together, one may ask, what their ranking is in terms of flexibility for human use. It seems that societal time is the most flexible of the four. It is infinite, and this is its most important asset. Being abstract and without any visible boundaries has made societal time a dimension that has been undergoing major changes since the industrial revolution. These changes will be traced in our discussion of time versus space within the urban context. Second in declining order of flexibility is personal space in its wider context of activity space. Although space is finite and bounded, it is still large enough compared to the size of a human being. If proper and cheaply priced communications and transportation technologies become available, then it will be possible for individuals to reach any point on earth or to enjoy maximum spatial flexibility, with but a limited time devoted to the moving process from one point to the other.

Third in order is societal space. Society may expand into new areas, especially within the urban context, but several limits would apply, especially the economic price of building new outlying facilities. Fourth, and most limited, is individual time, which is restricted by the finite human lifespan. This finitude is coupled with capability constraints of the human being that require a person to devote much time to physiological needs.

When it comes to societal time and space, several statements may be made that will serve as starting points for later discussions.

1. Societal time and space are more than the aggregate time and space of all the individual members within a given society. Society acts not only through individuals but through institutions as well. Then, too, individuals may operate beyond their immediate time and space. An individual may influence events beyond his/her lifetime. In addition, an individual may, through the use of communications, act beyond the activity space which he or she bodily reaches.

2. Societal time and space (as well as individual) are more than just limiting movement resources, as implied by Hagerstrand. Above all, they are economic resources and organizational-ordering dimensions.

3. At the societal level, there are constraints in addition to or instead of some of those identified by Hagerstrand. These relate to cultural values pertaining to time and space and to differences between men and women.

4. There are, at least at the societal level, positive enabling forces regarding the use of time and space. These may be related to the capitalist system or to national values, which call for a certain pace and certain norms of use of both time and space.

NOTES TO CHAPTER 1

1. Georgescu-Roegen, 1971, p. 130.
2. Ad Hoc Committee on Geography, 1965, p. 1.
3. Thrift 1977b, p. 448.
4. Hagerstrand, 1970; 1973; 1975; Chapin, 1974.
5. Pred, 1977a, p. 98.
6. Taylor and parkes, 1975.
7. A temporal perspective is typical to the study of diffusion, which from the outset has dealt with time and space simultaneously. The conception of time in diffusion studies has been criticized as referring to the use of time "as a ladder for processes to climb on" (Thrift 1977a), rather than adopting a chronosophical approach (Fraser, 1968).
8. Wilson, 1955; Carnap, 1958.
9. Harvey, 1969, p. 215.
10. Berry, 1964.
11. Harvey, op. cit.
12. Hagerstrand, 1973, p. 73.
13. Pred, 1976, p. 73.
14. The differentiation among four levels in symbolic logic was originally proposed by Carnap (1958).
15. The rationale for a locational (positivist) study of time-space at the macro-level of urban areas is that the nature of change in urban space can only be understood if a contextual study of both the temporal and spatial dimensions is considered simultaneously with the compositional changes in several economic and social sectors (see Kellerman, 1981). This perspective was demonstrated in a study of suburbanization and exurbanization trends in Philadelphia (Kellerman and Krakover, 1986).
16. Kellerman, 1987a; 1988.
17. Giddens, 1979; 1981; 1984.
18. For discussions and summaries of structuration theory see e.g. Giddens, op. cit., especially 1984; Pred, 1983; 1984; Gregson, 1987; Thrift, 1985.
19. The role of structuration as a contextual approach was recently discussed by Thrift (1985) and by Rose (1987). Several opinions on the importance of space to current social theory have been expressed in Urry and Gregory, 1985; Saunders, 1986, pp. 278-279; 285. See also Kellerman, 1987a.
20. This last point on humanism and structures in structuration was discussed by Pred, 1983.
21. General accounts on geography and structuration have been provided by Thrift (1983), Gregory (1982a; 1982b), and Pred (1982; 1984).
22. Pred, 1983.
23. Gregory, 1982a.
24. Gregson, 1986.
25. On bracketing see Giddens, 1979, pp. 80-81. It was implemented recently by Moos and Dear, 1986; Dear and Moos, 1986. Society at large as the preferred level of analysis was mentioned by Giddens, 1982, p. 66.

26. Gregory, 1978, pp. 88-89.
27. Harvey, 1984, p. 1.
28. Parkes and Thrift, 1980.
29. Goldman, 1978, p. 71.
30. Major works on the sociology of time include Moore, 1963; Gurvitch, 1964; Zentner, 1966; Kolaja, 1969; Fraser, 1975; Zerubavel, 1881.
31. Baker, 1981, p. 441.
32. Durkheim, 1915. On the distinction between private time and public time see Thrift, 1981; Kern, 1983, pp. 19-20; 33-34.
33. Luhmann, 1982, p. 303.
34. Bergson, 1959.
35. Giddens, 1979, p. 203; 1981, p. 42.
36. Giddens, 1981, pp. 44-46. See also 1979, p. 224.
37. Parkes and Thrift, 1980, p. 144.
38. Soule, 1955; Becker, 1965; Ghez and Becker, 1975.
39. Parkes and Thrift, 1975, p. 660.
40. Thrift, 1977a, p. 93.
41. Shackle, 1978; Lynch 1972; 1976.
42. Thrift, 1977b.
43. Ellegard, et al., 1977, p. 127.
44. Chapin, 1974; Pred, 1978.
45. On the broader societal future horizon, see Kellerman, 1981. On the longer societal past experiences see Giddens, 1981, p. 34.
46. Melbin, 1978a; 1978b.
47. Melbin, 1978a, pp. 5-6.
48. Rose, 1979.
49. Parkes, 1971; 1973; 1974; Thrift, 1977a.
50. Thrift, op. cit., p. 66.
51. For a detailed discussion of this point, see Kellerman and Krakover, 1986.
52. Curry, 1978, p. 44.
53. Carlstein et al., 1978, p. 3.
54. Carlstein's (1978, p. 154) absolute and relative times carry different connotations from those proposed in the preceding section on "time and space: context and composition" and in Figure 1. While Carlstein relates to social constraints and uses of time, my distinction in the preceding section related to philosophical aspects of time as context and composition. It is in line with the conventional distinction between absolute and relative space. This is one of several examples that will be noted of terms being used in different ways by various authors.
55. Hagerstrand, 1973.
56. Hagerstrand, 1975.
57. Carlstein, 1982, p. 58; Giddens, 1984, pp. 116-117; Storper, 1985, p. 409.
58. Gould, 1981.
59. Giddens, 1984, p. 177.
60. Rose, 1977.
61. Gold, 1980, p. 80.
62. Porteous, 1977, pp. 28-30.

63. Horton and Reynolds, 1971, p. 86.
64. Jakle et al., 1976, p. 92.
65. Ley, 1983, p. 102.
66. Hartshorn, 1950; Gottmann, 1952; Knight, 1982.
67. Soja, 1980.
68. Kellerman, 1985b.
69. Kellerman, 1981; 1985a; Kellerman and Krakover, 1986.
70. For humanistic-cultural geography discussions, see e.g. Relph, 1976; Tuan, 1974. For an architectural perspective see e.g., Lynch, 1960.
71. Israel, as a case study of a small country in which regional boundaries are sometimes blurred in spatial experiences, is discussed in Shamai and Kellerman, 1985.
72. Kellerman, 1981.

CHAPTER 2

TIME-SPACE HOMOLOGY: A SOCIETAL-GEOGRAPHICAL PERSPECTIVE

"Space is in its very nature temporal and time spatial."[1] This idea of a time-space homology, carried over from physics, has implicitly served as a point of departure for the recent development of time-space study. Time-space approaches using different perspectives, have been proposed for several areas of study. Thus, Harvey noted a general positivistic theory in geography: "it will explore the links between indigenous theories of spatial form and derivative theories of temporal process. The links run in both directions." On the other hand, Castells' radical perspective suggests that "from a social point of view [therefore], there is no space (a physical quantity, yet an abstract entity qua practice), but an historically defined space-time. a space constructed, worked, practised by social relations." By the same token, according to Soja, "it becomes less possible to separate history and geography, time and space.[2]

The purpose of this chapter is to present a framework and terminology for time-space study at the macro-societal level. The connecting thread between the discussion of time-space in general, on the one hand, and the several particular terms, on the other, will be the question whether there exists a time-space homology at the macro-societal scale. This is a basic question, since a hidden assumption behind the time-space analysis of individuals is that such a homology does exist. It will be argued that this cannot automatically be assumed for the societal-macro scale. Special attention will be given to three time-space homologues for the macro level: timing/ spacing, temporalization/spatialization, and temporality/spatiality.

The analysis of a societal time-space homology is important, since it could serve as yet another building block for the currently evolving social theory of space. It could illuminate the extent to which time and space may be considered homologous at the macro level. Furthermore, the interrelationships of these two basic elements of human life and social structure may be better understood if the specific time-space dual concepts are carefully applied to the societal level.

The chapter will start with a discussion of time-space at the societal level. First, the roots for the joint treatment of time and space will be highlighted, to be follwoed by an introduction of a simple typology of time, space and society at the urban level. This discussion will then be complemented with separate presentations of the possible application of three specific time-space dual concepts (timing/spacing, temporality/spatiality, temporalization / spatialization) to the macro level.

Approaches to Time-Space at the Societal Level

After putting space and time separately into a societal context in the preceding chapter, we may turn now to societal time-space as a homologous entity. With the exclusion of Hagerstrand's and Giddens' works, explicit discussions on time-space at the societal urban, regional, or national levels are relatively rare. The comprehensive text by Parkes and Thrift published in 1980, provided a detailed discussion of social time but very little, if at all, explicit discussion of time-space at the societal level. The use of the phrase "time-space" is itself problematic. It implies, a priori, that time and space are one entity or at least homologous. It is difficult to adopt such a general homology, since it has to be proved separately for distinct historical periods, geographical locations, and social levels. In a rarely quoted paper, Ed Ullman argued that "in general terms, space is conceived of as a passive and, to this writer, a more concrete dimension than time; time is a more active and more mental construct. Space implies being, time implies becoming."[3] Urry also made a fine distinction: "Although spatial change necessarily involves temporal changes, temporal change does not necessarily involve spatial change. Thus, we should distinguish between 'temporal' and the 'spatial-temporal', rather than simply the temporal and the spatial."[4]

On the other hand, one may assume some connection between time and space, sometimes at a very basic cultural level. This can be shown from perspectives as varied as language, human experience, religion, daily practice, history, philosophy, and social conditions. These varied perspectives on the connections between time and space will be briefly reviewed in the following paragraphs.

La Gory and Pipkin refer linguistically, to the fact that many non-spatial variables, including time, are interpreted in spatial ways. Giddens attributes this to the notion that all time measurements involve movements in space.[5] Experientially, it might well be that space, being the most visible element in primitive societies, influenced the expression of many other phenomena, especially time, which involves duration and is thus similar to space. From a religious point of view, it is interesting to note that the very first verses of the Judaeo-Christian tradition, namely the Creation story in the Book of Genesis, tell of the creation of both space and time on the first day (space first, time second).

On the more practical level of measurement, absolute space is measured in spatial units, while relative space (distance) is measured in temporal units, as well.[6] Historically, time and space were not always "same" or similar. Thus, Mumford argued that each culture has its own time and space. Thus whereas the Middle Ages were characterized by time and space being independent from each other, the Renaissance established the philosophical unification of time and space.[7] Heller refined this argument by noting that "the Renaissance concept of space had already become de-anthropologized, while its conception of time remained to the very end related to man." Moreover, space (Raum) "never appears in connection with time; the Renaissance recognized a time-space correlation only between time and space or position (Ort)."[8] On a philosophical level, Newtonian physics and the philosophy of science have shown that natural time and space are similar if not the same. Finally, on a social level, Hagerstrand showed in his writings that both time and space create constraints on an individual's movement that justify their joint treatment. All these aspects -- linguistic measurement, philosophical, social -- might explain some of the roots for the joint treatment of time and space. As will be shown later however, they do not provide any automatic justification for combining the two dimensions together when it comes to a societal analysis of modern societies.

Among the few who related to <u>societal</u> time-space is Carlstein, who referred to the time-geography of agricultural societies.)[9] Usually, such a view is ignored in societal studies, since the conceptions of space, when mentioned, relate to social and cultural spaces. This situation is unfortunate because time-space analysis easily lends itself to practically all of the currently prevailing modes of thought within geography. From a positivistic viewpoint one may extend the time-geography logic when looking, for example, for "the impact of a spatial economic change (pacemaker) on social-spatial organization, or how much time passes between a spatial-economic change until the spatial-political system adjusts itself to such a change (marker), and what are the spatio-temporal characteristics of such processes (synchronization, rate, duration, phase and phase shifts, limits)."[10] In terms of Marxist epistemology, "studies of people's control of time should encompass the class struggle, identifying the policies and powers of conflicting social groups."[11] An experiential approach would be able to study the impact of national-ethnic experiences and values on the societal perception of time-space, and on the manipulation and behavior in time-space.[12] Analyses of time-space at the societal level could provide excellent examples of the fact that these three approaches -- positivistic, experiential, Marxist -- may be complementary. In fact, the complementarity of the three epistemologies is revealed when a research problem is being focused on rather than an epistemological aspect, around which, of course, the extreme differences appear. The importance of blending several approaches has been stressed in both geography and sociology.[13]

Time-Space and Urban Studies

Following the exploration of the very idea of time-space at the societal level, we may now place side by side the three dimensions of time, space, and society. This comparison will identify some of their basic characteristics, and, by their integrative analysis, emphasize the special role of structuration. The discussion will focus on the city as a major object of social study along both time and space. A simple scale-classification of the three dimensions, as in Table 1, yields micro, meso, and macro levels of reference. Urban studies normally deal with one or more of these dimensions at the meso level; namely, with a city and its urban society along a period of several years. It is possible, of course, to apply micro or macro time to the study of a city from a spatial or societal viewpoint (e.g., long range trends on the one hand or urban daily life, on the other).

Table 1:   Scales of Time, Space, and Society

Scale	Dimension		
	Time	Space	Society
Micro	Hours, days	Room, House, Neighborhood	Individual, Family Small community
Meso	Years	City	Urban society
Macro	Decades, Centuries	Region, Country	Nation, Minority

The study of societal space and time involves, however, another basic differentiation -- the syntactic one between absolute and relative space and time. Absolute space and time are passive or "automatic" from a human viewpoint, and they serve as containers for social life. Relative time and space only become active dimensions through human conception and action; otherwise, they may be viewed as movement, or economic or political resources. Their use at the societal level depends on culture, ideology, technology, and socio-political organization.

Combining scale and syntax differentiations permits a comparative view of several disciplines focusing on the city (Table 2). The study of relative-active time and space of urban society at the meso level (urban societies in cities through the years) is handled by a structurationist approach. With the use of this approach, changing conceptions and uses of time and space by urban societies are at the core of interest. However, the study of the conceptions and uses of both time and space should not imply that whatever is true for time is also true for space, or vice versa.

When presented graphically (Figure 2, following a similar approach by Holly for the individual-micro level), the most studied inter-connections among time, space, and people seem to have taken place at one or more of the extreme edges of the three continuous dimensions. Relatively little attention has been paid to the focus of the three, or the joint meso level of urban society/city/years. Thus, for example, behavioral geography has focused on small spatial human units, while urban planning has dealt, at least partially, with a relatively distant future for city and region.

The time dimension lends itself to two spectral presentations in contrast to space and society, which range mostly along size. One, which was used here, is along tense, from past to future; the other is from short/small time units (minutes, hours) to long/big ones (decades, centuries). The first option was preferred here, since time units are less central than is time tense when this dimension is compared to space and society. Furthermore, time has been treated mostly as a passive than an active dimension, so that its direction is more significant than its size. We shall return to time size both with a discussion of temporality, and in Chapter 5, where Israeli temporalities will be discussed.

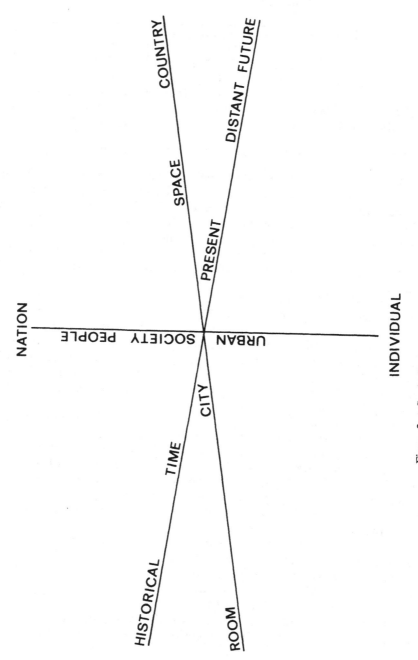

Figure 2: Study Scales of Time, Space, and People

Table 2:   Time, Space, and Society in Some Urban Studies

Approach	Dimension					
	Time			Space		Society
	|Scale	Syntax	Tense	|Scale	Syntax	|Scale
Historical Geography	|Macro	Passive	Past	|Meso or |Macro	Passive	|Meso or |Macro
Time-Geography	|Micro	Active and Passive	Past or Present	|Micro or |Meso	Active	|Micro
Urban Ecology	|Meso |or |Macro	Passive	Past or Present	|Meso or |Macro	Passive or Active	Meso or |Macro
Planning	|Meso |or |Macro	Passive	Future	|Meso or |Macro	Active	|Meso or |Macro
Structuration	|Micro, |Meso, |or |Macro	Passive and Active	Past, Present or Future	|Meso or |Macro	Passive and Active	|Meso or |Macro

Giddens' Approach to Societal Time-Space

An exception to the rarity of explicit discussions of time-space at the societal level are Giddens' writings.[14] Three major aspects of his approach to time-space at the societal level, will be reviewed here: the nature of time-space, time-space distanciation, and time-space and urbanism.[15] We shall concentrate here on the nature of time-space and the differences between time and space; the analysis will put into question whether time-space constitutes a homology, at least in Giddens' writings.   Giddens rejected the view of time-space as mere "environments," "containers" or "categories of mind." He preferred the view that "time-space relations are portrayed as constitutive features of social systems, implicated as deeply in the most stable forms of social life as in those subject to the most extreme or radical modes of change." Based on Heidegger and Leibnitz, however, "they [time and space] <u>are</u> the modes in which relations between objects and events are expressed."[16] These two views are not the same.

The first refers to time and space as compositional elements, while the second relates to time-space as contextual dimensions. Giddens has not made it known whether his two views are two contradictory alternatives or whether they are complementary. His observations/terms, however, relate to both contextual and compositional time-space. Thus, time-space distanciation is usually contextual, ("the media of time-space distanciation")[17] while the commodification of time and space and the notion of authoritative resources are compositional.[18] Moreover, it is not clear whether Giddens thinks that time-space constitutes a homology in a given era and society; in other words, do they share similar modes of expansion and usage, are they distinctly separate elements with a different and changing importance along a given period and society? Interesting, too, is that Giddens did not provide a justification for the joint treatment of time and space at the societal level as their being "paths involving collectivities rather than individuals."[19]

Time-space distanciation relates to the fact that, as Giddens put it, every social system in some way 'stretches' across time and space."[20] This definition seems to be close to a "container" absolute-passive view of time-space, especially when it is related to societal "storage capacity." More important, however, are the hidden assumptions in this definition that when there is a distanciation of time, it is automatically coupled with a distanciation of space, and that both processes are operating at the same pace. The use of Giddens' major examples may assist in demonstrating this criticism: Before the innovation of irrigation, time was mainly experiential (which is another view of time) and restricted to a close past. Irrigation made it necessary to plan for a future time but, again, for a relatively short period of time. Spatially, the extent of distanciation became relatively large when more areas came under cultivation. The contrary is true for the introduction of writing. Although it assisted in controlling a distanciated space, the major impact of writing was on time in causing a major distanciation to past and future (i.e., making them relatively long). In any case, Giddens views time and space in this regard as socially enabling forces rather than with the constraining outlook proposed by Hagerstrand for individuals. Time and space are crucial elements in the structuration process, and therefore Carlstein would like to see more discussion on the time-space constraints of structuration.[21]

Urban capitalism is characterized by Giddens as, among other things, commodifying time and space.[22] The commodification of time was done in the capitalist work place; that of space was achieved through "creative space" although that already existed in class-society.[23] Space had been commodified very early in human history. One of the earliest accounts is Abraham's purchase of a gravesite for his wife, Sarah, as described in the Book of Genesis. Capitalism, thus, intensified the already existing commodification of space. As we shall see in the next chapters, the process of time commodification in capitalism has made time into a more important resource and dimension than space, although the latter had been a commodity many years before time became such.

Despite the somehow ambiguous connotation of time-space in Giddens' structuration, his writings are pioneering in societal time-space from a geographical viewpoint. His concepts, especially distanciation, represent attempts to understand when and through which processes the social becomes spatial and temporal (though he does not refer to the process explicitly in these terms). This transformation and the other side of the coin -- namely, how the spatial and temporal become social -- may be studied through the concepts of temporality and spatiality, which will be done in a later section.

Time, Space, and Society at the Urban Level

The general discussion thus far presented may lead one to conclude that time and space at the urban societal level are both passive and active dimensions. Urban society expands and extends its spatial and temporal dimensions by further invasion into time and space. On the more active side, and simultaneously with the expansion into passive time and space, urban society uses time and space as production resources. Lefebvre's view of space as becoming a social product may be extended to include time, as well, since the use of time also reflects social values and structures.[24] Thus, urban time and space are both production resources and social products of capitalist urban societies. The level of expansion into time and space and their use by urban societies change along time. Developments in communications and transportation technologies permit the use of more space for urban areas and make time use faster, more instant, and more efficient.[25] Structural economic changes, especially the move from an industrial economic accent to services and information activities, also contribute to the expansion into time and space.[26] Footloose facilities, as many of the service and information activities are, call for further urban expansion in space. The ability to receive and use information instantly through the use of computer technologies may change attitudes toward time pace and use not only in production but in urban life in general.

The uses of time and space and the further expansion into them do not necessarily have to be at the same pace and pattern. Telecommunication technologies may permit the use of more urban space while using less time for both production and consumption. Time and space are, thus, only partial homologues at the urban societal level. The common phrase, "time-space", cannot be used in a rather general sense, and the connotation of this term has to be specified.

A general dialectic relationship pertains in the structuration process between human action in time-space and time-space structures.[27] This relationship applies especially at the urban level where both expansion into time-space and further use of these two elements are ongoing processes. Individuals may, for example, move their residences farther outward in an urban area, assuming that a basic infrastructure exists to permit this move. The aggregated residential move of many individuals creates a new societal conception and use of urban space, since the urban area is now larger and might as well have new spatial and social patterns. The attitude toward time and its use might change, too, as a result of urban spatial growth, since larger distances require further time use. The very act of moving outwards is itself dependent on social norms and values with regard to the use of urban space (or time); if society does not encourage further urban expansion, its likelihood of occurrence will be low.

A pioneering attempt at urban modeling of structuration is the work of Moos and Dear.[28] Their model details the several participants, aspects, and levels of an urban structurationist process. The duality of structure that is the heart of structuration is left in a black box. As far as time and space are concerned, it seems that the two dimensions are treated contextually in the model, though Moos and Dear were aware of their compositional importance, as well.

The Three Time-Space Homologues

The preceding sections dealt with the general concept of time-space. The study of individuals and society in time-space has produced three specific homologues: timing/spacing, temporality/spatiality and temporalization/spatialization. Their possible uses and significance for a societal time-space framework will now be discussed. It is important to draw attention to these homologues, especially spatiality/temporality, since the use of the phrase "time-space" is too general and not always justified. It is equivalent in its generality to using the term "society" for every social phenomenon, ranging from socializing to social structure, and to using the general term "space" for every geographical aspect and process.

Timing/Spacing

The dual-concept of timing-space and spacing-time was proposed by Parkes and Thrift for individual rather than societal time-space.[29] It may be partially extended to societal time-space, as well, however. This dual-concept is not a simple homologue, since timing-space does not carry the same connotation as spacing-time. Timing-space was defined as the use of "time as a means to the patterning of space", while spacing-time was "the process of ordering objective (clock or calendar) time which is available for allocation to events and the awareness of the associated objective and subjective durations of those events, and their linkage to other events.[30] In defining the timing of space, Parkes and Thrift thus referred to space in its geographical sense; while their definition of spaced-time relates to space as a temporal interval.

This dual-concept may be extended, though very partially, to societal time-space. Of the seven different ways in which space is timed according to Parkes and Thrift, only two apply to societal time-space:[31]

> Space is timed by the <u>duration</u> of activities which occur in a given space and indirectly by the location of an activity <u>X</u> within the <u>sequence</u> of activities (a, b, c, ..., x, ..., z). Space is timed in terms of the number of non-redundant (that is, synchronous and synchronistic) activities that occur over some time period t_0-t_1. In this sense we may conceptualize city-fast and city-slow areas. Such areas may be functionally similar, for example both may be

residential areas, or both may be retail areas. There is an opportunity for policy-related intervention when the markers and pacemakers of time-space have been isolated.

In the first option, which deals with <u>duration</u>, space is passive, and only time is active. In the second option, which deals with <u>pace</u>, space and time are both active and passive. Thus, different urban spaces may be used as resources in several ways, and these spaces may shrink or expand during or as a result of changing uses. Time is passive in the sense that urban changes occur in it; but it is also an active resource that determines the pace of these changes. The concept of timing-space may, thus, be extended from a time-space comparison at the intra-urban level to inter-city and inter-regional comparisons.

The patterns of timed spaces with regard to both duration and pace are determined by the aggregate activities of many individuals and institutions, which, on their part, are integrated in deeper social structures and cultural superstructures. On the other hand, the activities of both individuals and institutions may gradually change existing structures and superstructures. Time and space are, therefore, basic intermediate forces in the structuration process, in this case in the form of timing-spaces and timed spaces. Once again, time and space are seen, as being simultaneously passive and active social elements. It would be artificial to ignore the role of either time or space or to relate to them only as active <u>or</u> passive. This is especially true when duration and pace are examined. Communications may serve as catalysts in the changing patterns of pace and duration of time and space. The latter, on their part, may determine human action at both the individual and institutional level.

Spacing time at the societal level is even more limited than timing-space when the several scenarios for the time-spacing of individuals are examined. Thus, only the following scenario may be applied at the societal level, as well.[32]

> Spacing of time may also occur indirectly as the result of the relocation of activities in geographical space, thus it is not always an "active" process but may be quite "passive." When events change their geographical location they will always cause a respacing of time which extends beyond those in question, that is they have been relocated. In the initial ante-relocation environment the sequence of event participation may have been quite different.

Here, too, both time and space are both active and passive elements at the same time. Activities move in space and cause an extensive spacing of time. Time and space are, again, two-way intermediate forces between structures and superstructures, on the one hand, and human action, on the other.

Temporality/Spatiality

Compared to the somewhat restricted sense and use of timing-space and spacing-time at the societal level, temporality and spatiality were developed within a societal framework. Their application is more general, carrying the same connotation for both temporality and spatiality and, hence, turning them into a full conceptual homologue.

It is difficult to point to a strictly formal definition of spatiality and temporality in the literature although the terms have been mentioned by several writers.[33] Soja did provide explicit discussion of spatiality, emphasizing the distinction between <u>contextual space</u> and <u>created space</u>. The latter term, originally proposed by Lefebvre and Harvey, is at the center of Soja's spatiality.[34] Space as a created entity is thus seen as a <u>social product</u>. Created space is one dimension (the social) of the more general <u>compositional space</u> that has been proposed as a counterpart to contextual space. Spatiality according to Soja, is another term for "the spatial organization of society." The strong social connotation of spatiality is expressed in its definition: "...a social product which is created, shaped and transformed by the same structural forces, antagonistic social relations, and periodic crises and struggles which affect the production process and social life more generally." Moreover, as a social product, "spatiality is also a material force which reflects back upon these social processes."[35] This definition is also the essence of Soja's "socio-spatial dialectic." Although space and spatiality are interpreted as social products and as reflecting on social processes, Soja's spatiality still differentiates between the "social" and the "spatial". What is missing in order to facilitate the dialectic between society and space is the <u>societal conception of space</u>. Is space viewed merely as a container or is it also a resource? Is it more of a social or political resource or is it rather an economic resource or is it perhaps both? These differentiations are still very crude when a specific society is the focus of study. One may assume that a particular conception of space reflects a unique mix of political, cultural, social, and econmoic aspects, in addition to the impact of an existing spatial organization of society. Obviously, these observations on space and spatiality are also true for time and temporality.

The concept of temporality was addressed explicitly by Gross, who defined it as "the span of historically interpreted time, usually stretching from some originary point in the past (a beginning, a founding movement) down to the period of one's lifetime."[36] This form of temporality Gross also terms <u>longue duree</u>, which differs from two shorter durations, the day-to-day one and the one relating to biological lifespan. Temporality is thus synonymous with memory. Several comments have to be made regarding Gross' definition. First, <u>longue duree</u> is not only <u>longer</u> than the two other forms of temporality, but foremost, it is also <u>larger</u> in the sense that its content relates not to one's individual but to societal-collective experiences. As such, it has wider significance regarding both present collective or shared uses of time and societal expectations of the future. Second, the emphasis on time remembrance and interpretation excludes time use, thus giving temporality a rather more passive connotation. Third, Gross has shown the impact of historical time-memory on society and <u>vice</u>

versa, but this crucial two-way relationship has not found its way into his definition of temporality. Finally, he proposed three stages for changing temporalities in modern Western society: the religious, the national, and the capitalist. We shall return to these stages in our discussion of Israeli society (Chapter 5).

The two concepts, spatiality and temporality, may be defined as the conception and use of time and space by society (or individuals, on a different level). The conception of space and time is the process by which the spatial and temporal become social, since existing spatial and temporal patterns and uses shape social values and norms regarding time and space. The use of time and space is the process by which the social becomes spatial and temporal, since these uses reflect social structures and values. Spatiality and temporality, therefore, are terms that encompass the two-way transformation processes between society, on the one hand, and time and space, on the other. As such, spatiality and temporality are basic elements of structuration. The idea of societal spatiality and temporality may originate with Kant, who viewed time and space as "pure forms of sensible intuition."[37] Time and space are matters of individual or societal conceptions or, to paraphrase Lefebvre, the concepts of time and space are not within time and space but in Man and society.[38]

The meaning of temporality and spatiality as the "use of time and space" refers to whether time and space are used extensively or intensively and the extent to which society "expands" into time and space. With this last point, temporality and spatiality seem to be closer to Giddens' "distanciation in time and space."[39] Two points should be mentioned in this regard, however. First, societal temporality and spatiality are wider concepts than distanciation, since they refer to the conception of time and space as well as to their use. Second, distanciation seems to assume, at least implicitly, similar "expansion" patterns into both space and time. It will be argued later that spatiality and temporality do not necessarily share the same patterns within capitalist urban societies.

The dual concept temporality/spatiality makes it possible to reify time and space and their meanings through their societal uses. Space is a physical entity itself, while time is an abstract dimension that cannot be considered apart from its contents.[40] Still, even space per se does not make any societal sense beyond its geometrical properties, unless certain significances are attached to it. Merely stating that the universe exists by definition in time and space does not reveal the societal meanings and uses of the two dimensions for society. Since it is almost impossible to separate time and space from their uses and connotations, it is through these properties that societal time and space may be reified -- and hence the basic need to use the terms spatiality and temporality in the social study of time and space. In short, temporality and spatiality provide a bridge between abstract time and space and the objects and events through which they are revealed and become socially meaningful.

Structuration is an important language and construct in the study of spatiality and temporality, since it assumes a constant dialogue between society and social values, on the one hand, and time and space, on the other. Time and space are, thus, passive in the sense that societal action takes place in them; but they are also "active" in the sense that they may be used as active resources and dimensions. The use of time and space is bounded by social values, but these uses may in return reshape those same.

Very often, therefore, studies of societal time and space are, actually studies of temporality and spatiality. Althusser and Balibar assigned different temporalities to several societal levels, while Lipietz did the same for spatialities;[41] these writers, though, were more concerned with pace than with the conception and use of time-space. Temporality and spatiality in the context of the conception and use of time and space may be applied to many social studies. From a spatial-geographical viewpoint, an interesting avenue could, for example, be the study of differences in the conception and use of time and space between urban and rural societies. From a temporal viewpoint, differences in spatiality and temporality with regard to work and leisure time could be of interest, particularly since work and leisure are differentially related to by different social groups.[42] From a societal perspective, it should be of interest to study, for example, differences in spatiality and temporality between working mothers and other workers regarding commuting.

Spatiality and temporality at the societal level (bourgeois society between the late eighteenth and early twentieth centuries) have been referred to as the perception of space and time.[43] This perception has been reflected in changing artistic styles and other cultural expressions. It is sometimes difficult and even artificial to differentiate clearly between perception and conception at the societal level. Societal perception of time and space carries, however, a more experiential and passive connotation, which finds its expression in art, architecture, and culture. Societal conception of time and space carries a more socio-economic and active connotation. As such, it finds its expression in production and consumption modes and in social and spatial organization.

Temporality and spatiality may be applied to time-space studies at the individual level, as well, since the movement of individuals in time-space constitutes their spatiality and temporality. The term "action-space," which refers to the geographical action field of an individual is, thus, part of a person's spatiality. By the same token, an equivalent "action time" would be part of a person's temporality, referring to his/her active hours or days. Time-geography emphasizes the contextual rather than the compositional, so that spatiality and temporality within it are only partial and the conception of time and space by individuals is excluded. Individual time-space is beyond the scope of this volume; hence, it will suffice to note that integrating the knowledge on spatial conception gained in behavioral geography with time-geography could potentially produce a fuller temporality and spatiality for individuals.

Spatiality and temporality present a homology that could have interesting implications for the evolving common field of sociology and geography. Both are social terms, since they relate to people's conception and action in time and space. On the other hand, people's uses of time and space result in the spatial organization of society that is of major interest for geographers. Comparing the temporalities and spatialities of several nations may shed new light on different modes of spatial and temporal organization. Although temporality and spatiality constitute a conceptual homologue, this cannot lead to the more practical assumption that the conception and use of time and of space within a certain society will have similar patterns of pace and intensity. Time may be used in an intensive manner and space may be in extensive modes or vice versa. Thus, a homologous definition of time and space concepts does not automatically call for a practical, homologous, societal use of the two, as well. This last assertion will be the main thesis of the next chapter.

Temporalization/Spatialization

A third duo of time-space concepts is temporalization/spatialization, which were introduced by Gross as being recent and strong social processes.[44] Gross defined spatialization as "the tendency to condense time relations -- which are an essential ingredient for personal and social meaning -- into space relations. Nearly everything that should be understood through categories of continuity and duration is now approached through categories of space and measurement. One whole dimension of life is not only being defined in terms of another, but is being reduced to it. Hence, vertical relations are made to collapse into horizontal ones."[45] The signs of spatialization include an urge for immediacy in modern life, spatial rather than temporal identities and correspondencies in comparisons, and a non-historical consciousness. Among the major causes of spatialization Gross counts urbanization and technology. There is not only a spatialization of time, according to Gross, but also of culture and the social sciences.

Although Gross has not provided a formal definition of temporalization, it may be understood implicitly to mean a mode of thought (which can be achieved only by intuition and introspection) that would temporalize space, culture and social thought. Thus, an understanding of space would have required a stronger emphasis of continuity and duration rather than measurement, and vertical relations would play a larger role than horizontal ones in spatial analysis. The two concepts of spatialization and temporalization are, thus, homologous. Gross emphasized spatialization rather than (the more preferred) temporalization, since this is the currently prevailing process.

The concept of spatialization does not have the simple connotation of "putting things in space," since it refers to the measuring, thinking, and conceiving of time in a "space-like" manner. Therefore, spatialization does not necessarily lead to the predominance of spatiality over temporality. In other words, if time is perceived, studied, and even used in forms traditionally attributed to space, it does not mean that space is more important than time in social life. On the contrary, it might well be argued (as it will be in the next chapter) that the spatialization of time has given the latter the once immense powers of the former. And the same would be true of the temporalization of space. By the same token, the spatialization of time in thought does not automatically lead to a complete spatialization of temporality in all facets of life. This would mean that time not only is conceived in spatial modes but also is so used. As will be shown later, the opposite is true. Similarly, Gross does not show that the current trend of time intensification stems from space or spatiality; thus, it is difficult to call a quantification in the conception of time simply as spatialization. The whole issue becomes more complex if the current temporalization of space is also considered through the measurement of distance along temporal units.[46]

At the philosophical level, the spatialization of time was condemned by Bergson, whose arguments on this matter have been found similar to those of Proust and Freud, all three of Jewish heritage.[47] This stems, it has been suggested, from a heavier emphasis on time in Judaism than on space, a topic that will be discussed within the context of time and space in Israeli society.

Spatialization and temporalization are different from spacing and timing, since the interrelationships of time and space are not organized around some events but rather around social and cultural change. The time-space interdependence here is deeper; it involves processes at the superstructure level in that modes of thought and cultural values are changed. Spatialization and temporalization are related to spatiality and temporality, since spatialization transforms temporality and temporalization transforms spatiality. Furthermore, temporalization and spatialization may be interpreted as specific forms of spatiality and temporality, since they relate to certain conceptions and uses of time and space. The impact of individual human action on these transitions is in the long run comparable to more immediate personal impacts on timing/spacing and faster individual impacts on temporality/spatiality. The reason is that superstructural changes in cultural values, thought, language and science are slow. On the other hand, the societal impact of spatialization and temporalization may eventually be immense, as shown by Gross, with regard to social life, consumption, culture, thought, and experience.[48]

Conclusion

In the preceding sections, we have attempted to discuss the possible options for a time-space homology from a societal-geographical viewpoint and to illuminate them from a structurationist perspective. Time and space were interpreted as being both active and passive elements of urban societal life, so that society uses them as resources and simultaneously expands into them. The use and expansion processes do not necessarily have the same pace and duration in space and time at a given period; however, they both reflect the ongoing relationships between human agency and social structure. Timing and spacing emphasize the special roles of time and space as intermediate forces in this process. Temporality and spatiality reflect the changing societal conceptions and uses of time and space. Temporality and spatiality may also undergo mode-changes through spatialization and temporalization.

It has been shown that although time-space is usually referred to, at least implicitly, as a homology, it is not always one at the societal level. Careful attention has to be given to the differences between time and space at this level, especially when the three more specific time-space homologues are analyzed. Theoretical differences between the definitions of time and space could exist as is the case with resepct to timing/spacing. In addition, there might also be differences at the practical-empirical level, as has been argued by Gross, concerning spatialization/temporalization. The most useful dual-concept for further study of time-space is spatiality/temporality because of the wide range of both its conceptual meaning and its applicability. It cannot be argued, however, that there will always be a homologous relationship between spatiality and temporality at the practical-empirical level. In other words, although spatiality and temporality share the same conceptual definition, a societal empirical analysis may show that spatiality is not the same as temporality in terms of modes, form, and pace.

This chapter has attempted to show that the strong, complete time-space homology for individuals does not necessarily exist for the societal level as well. This does not mean, however, that the hyphen between time and space should be completely removed in societal analysis. Rather, it calls for a more careful look at time and space, both contextually and compositionally. Time and space are basic social elements and are strongly related to social structure through the structuration process. Continued time-space analysis may, thus, shed further light on social life as a whole.

NOTES TO CHAPTER 2

1. Alexander, 1920.
2. Harvey, 1969, p. 129; Castells, 1977, p. 442; Soja, 1979, p. 3.
3. Ullman, 1974, p. 126.
4. Urry, 1985, p. 31.
5. La Gory and Pipkin, 1981, p. 110; Giddens, 1979, pp. 204-205.
6. Falk and Abler, 1980 p. 64.
7. Mumford, 1934, pp. 18-21.
8. Heller, 1978, p. 170.
9. Carlstein, 1978; 1982.
10. Kellerman, 1981, pp. 20-21.
11. Baker, 1981, p. 440.
12. Kern, 1983.
13. In geography see Morrill, 1984, p. 68. In sociology see Giddens, 1982, p. 66.
14. Giddens, 1979; 1981; 1984.
15. Giddens' approach to time-space has been influenced by Hagerstrand's time-geography, and has been reviewed from this and other viewpoints by Carlstein, 1981; Gross, 1982; Gregson, 1986.
16. Giddens, 1981, pp. 30-31.
17. ibid., p. 157
18. The commodification of time and space relates to the process by which the two change from having a use value to having an exchange value. See Giddens, 1981; 1984.
19. Giddens, 1979, p. 224.
20. Giddens, 1981, p. 4.
21. Carlstein, 1981.
22. Giddens, 1981, p. 149.
23. ibid., p. 146.
24. On space as becoming a social product, see Lefebvre, 1974.
25. See Kellerman, 1984.
26. Kellerman, 1985c.
27. It was noted in passing by Pred, 1983, p. 47.
28. Moos and Dear, 1986.
29. Parkes and Thrift, 1975.
30. ibid., p. 656.
31. ibid., p. 659.
32. ibid., p. 660.
33. E.g. Soja 1979; 1980; 1982; 1985; Gregory, 1978, p. 119; Althusser and Balibar, 1970, p. 99; Lipietz, 1975, p. 417; Gross, 1985.
34. Lefebvre, 1974; Harvey, 1973.
35. Soja, 1982, pp. 250-251.
36. Gross, 1985, pp. 53-54.
37. Kant, 1961, p. 67.
38. Lefebvre's (1974, p. 345) original phrase was on space only.

39. Giddens, 1981.
40. Piaget, 1971, pp. 60-61.
41. Althusser and Balibar, 1970; Lipietz, 1975.
42. Parker and Smith, 1976.
43. Lowe, 1982, pp. 35; 59.
44. Gross, 1981; 1982.Temporalization and spatialization were introduced earlier with a more limited sense of identity by Taylor, 1955.
45. Gross, 1981, p. 59.
46. See e.g. Falk and Abler, 1980.
47. Bergson, 1959, pp. 98-99; 220-221.   On the Jewish roots of the attitude to time of Bergson, Proust and Freud, see Kern, 1983, pp. 50-51.
48. Gross, 1981.

CHAPTER 3

TIME VERSUS SPACE IN MODERN URBAN SOCIETIES

At the societal-urban level, time and space may be used and referred to in inverse and contradictory trends. Lying at the heart of this argument is the significance of time and space for urban society, especially as production resources, as it is impossible to separate society from the space in which it lives.[1] The approach taken in this chapter is one that moves back and forth between society and time-space as the elements studied.

A structurationist framework for a geographical study of time and space has been recently applied to a theory of "place becoming".[2] That theory has a strong humanistic accent emphasizing personal perceptions and feelings for place, or the micro level. The arguments to be put forward here emphasize, instead, societal uses of time and space. As such, the focus is on people and institutions, or the macro level. Foremost attention will be given in this discussion to actions, or the uses of time and space. These uses are affected by social values, and in turn they influence societal conceptions of time and space. Actions and conceptions, therefore constantly undergo simultaneous change, and so should be viewed in light of social, economic and historical circumstances. This duality of structure implies that in the aggregate, people's uses of time and space participate in the creation of new conceptions of time and space. At the same time, the uses of time and space are influenced by existing conceptions. This dualism may be described both historically and for contemporary urban society, by comparing the structuration of time and space. Before we do so, however, several statements have to be made. Making use of the two dual-concepts of temporality/spatiality and temporalization/ spatialization, these statements will serve our analysis of societal attitude to time and space in an urban capitalist context. They are as follows:
1. Temporality and spatiality have developed since the industrial revolution in an inverse form; that is, temporality has become extremely intensive while spatiality has become extremely extensive.
2. These inversely related elements, temporality and spatiality, are interrelated through transportation and telecommunications technologies.
3. Spatial forces -- i.e., urban and regional planning -- have little impact on temporality, whereas time control, if it would have been possible, could have had an impact on spatiality.
4. Temporality and spatiality have undergone a full-cycle mutual change in that space has been temporalized and time spatialized.

This chapter will now devote itself to an elaboration of these statements. Major emphasis will be given to the first two, as they form the base of the latter two. We shall begin by treating changes in temporality and spatiality separately, then turn to comparative discussions of the two elements and, lastly, discuss the implications of our analysis.

Several other comments have to be made on the nature of the sections that follow. The arguments will be wide ranging, in terms of both the concepts employed and the time span covered. Many arguments will, thus be presented or proved through the literature that has developed in the social sciences and history. This kind of analysis requires the use of terminologies taken from several disciplines. Although some of the literature may be open to different interpretations, the accumulated evidence seems to highlight some general trends. Evidence will be presented for changes in both the concepts of time and space and in their uses. The discussions on transformations in temporality and spatiality will, obviously, be reductionistic in style and content. They should, though, highlight, "the spirit of the age."[3]

From Time to Time

A discussion of the societal significance of time may basically relate to two forms of time, namely historical time and human time. The first is of a "passive" nature, since events are normally supposed to flow on a temporal scale beyond the life span of an individual. This is the <u>longue duree</u>.[4] Human time, on the other hand, is the time used by individuals and institutions on a daily and annual basis, especially for planning, movement, work and leisure. In modern societies, time has a more "active" connotation since it is being used actively by people and institutions. Emphasis in our discussion will obviously be on human time although historical time will not be completely ignored. After all, one may argue that historical time, or the passage of events, may influence the use of time, especially in traditional societies.

The transitions from agricultural and town economies to urban industrial and post-industrial economies involved major changes in temporality. Very generally speaking, the attitude toward human time changed from relating to it as something passive that "passes by" (as does historical time) to thinking of it as an "active" thing, a resource and a basic dimension of societal-economic livelihood and prosperity. This important change has been studied or addressed in passing by scholars of different disciplines and adhering to various philosophies.[5] As a result, different names and terms have been proposed for the same processes and phenomena. The changes in temporality since the industrial revolution may be seen in Figure 3.

Until late in the eighteenth century, time was mainly related to <u>cyclically</u>, in the form of a circle, a view typical of traditional societies on a cross-cultural basis even nowadays.[6] Time was considered a continuous recurring rhythm through events such as seasons, birth, maturity, and decay.[7] This <u>cyclical</u> notion of time was usually coupled with what has been called a polychronic approach of people to time; that is, the performance of several functions simultaneously in a form that is rather casual and unstructured.[8] The users of such time, as Parkes and Thrift phrased it "are more heavily reliant on time relations in the present than in the future. Item or event order is imprecise and the duration of events is relatively flexible or elastic."[9] Thus, society was, at least from the perspective of a modern observer, in a "time surplus" condition.[10] Moreover, "as

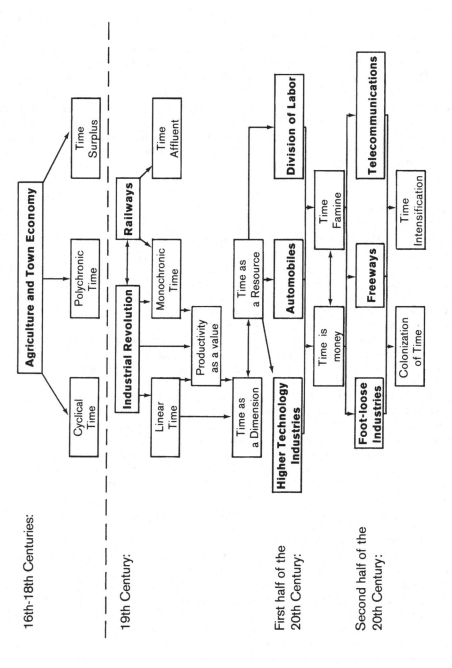

Figure 3: Time in Capitalist Urban Societies

long as power was concentrated in the ownership of land, time was felt to be plentiful and was associated with the unchanging cycle of the soil."[11]

Even under the dominance of cyclical time, another notion of time existed, that of <u>linear</u> time. In this concept, time, <u>in abstracto</u> and through events in it, was assumed to be something leading or "pushing" upwards to an endless future.[12] The cultural roots of linear time are difficult to pinpoint. Giddens proposed the development of writing, as a major <u>innovation</u>, to be "the first emergence of the 'linear time consciousness.'"[13] Whitrow argued that an <u>event</u>, specifically the Crucifixion -- "a <u>unique</u> event not subject to repetition" -- was the source of linear time.[14] Although Giddens, too, claimed that the emphasis on the non-repeatibility of events, was specifically Christian, one finds a much earlier event not subject to repetition: the Divine promise to Noah in Genesis that the deluge will not happen again.[15] (Ironically, the Bible connects this promise to the evolution of the four-seasoned year, which is cyclical time).

Early Christian roots for clock rather than natural-cyclical time may be found in the Roman church of the third century, and it received particular expression in the establishment of public time in monasteries.[16] In any case, European peoples lived by cyclical rather than linear time until the post-medieval era, when notions of linear time started to become more important.[17] Again, it is difficult to point at one specific or several definite causes for this change. To mention just a few factors in several areas: the interest in astronomy, puritanist morals, cultural-artistic developments, the invention of the mechanical clock, and Christian traditions. The Renaissance, too, contributed to a developed understanding of temporal concepts such as irreversibility, point in time, continuity of social events, and consciousness of this continuity and rhythm. The pace of time, consequently, became faster.[18]

The industrial revolution emphasized measured production time. Time became the basic medium and resource for the new industries and for the novel social value of productivity. The mechanical clock, and more so the popular use of hand watches, served as a stimulus for individualism though they marked public time. No wonder, then, that it was the British watch industry that flourished in the 18th century. The industrial revolution resulted, thus, in a societal change, from cyclical to linear time, the roots of which were much earlier. This change in temporality was not sudden. In 1754 it took two days to go from London to Bristol. Thirty years later it took only sixteen hours. The introduction of the first steam locomotive in 1820 came after fifty years of speed record-breaking.[19]

The view of time as linear was coupled with an additional change, from looking at time as something passive to viewing it as a resource: the so-called commodification of time.[20] The use of time moved from task orientation to timed labor, so that it was divided into owner's time (work) and own time (leisure).[21] Time, therefore, received an exchange rather than a use value. In parallel with these developments, time was becoming a separate entity from space, a quantifiable phenomenon in addition to a lived experience.[22] Time also became abstract, separated from nature, recorded, standardized, and made interchangeable.[23] On the cultural level, bourgeois society from the late eighteenth to the early twentieth centuries, began to experience time in new ways, no longer comparable to space. "With time, one order in space could be connected to another order in another space. Yet development was a new connection which posited dynamics (as opposed to statics); transformation (as opposed to unrelated, specific change); structure (as opposed to taxonomy); and totality (as a spatio-temporal whole)."[24]

The industrial revolution not only changed time conception and use, or temporality, it also caused new interrelationships to evolve, between the use of time as an "active" and important factor, on the one hand, and technology and socio-economic values on the other. The social demand for more efficient time-use brought about technological innovations to satisfy this demand, and these innovations on their part called for a more efficient use of time. Even the simple new mode of transportation, the bicycle, meant a traveling speed four times faster than walking.[25] It was left for the railway revolution in the first half of the nineteenth century however, to become the major step in this direction of greater efficiency. As Momford put it, "technologically, the department in which paleotechnic industry rose to its greatest eminence was not the cotton mill but the railroad system. The success of this new invention is all the more remarkable because so little of the earlier technique of the stage-coach could be carried over into the new means of transportation."[26] In addition to speed of movement, or as a result of it, the railway made it possible for people to be more punctual in their meetings and business affairs.

On the other hand, operating the railway system required time coordination among stations. These social and organizational needs led to the introduction of standard time, first on regional and national scales and then, toward the end of the century, on an international scale. The introduction of standard time was aided by the invention of the telegraph in 1837.[27] By mid-century, railways had a speed of 40 miles per hour, cities grew (as we shall see later), and increased diversity and specialization in economic activities required more precise temporal coordination.[28]

Another temporal change in the nineteenth century was the social conception of a more complex present because of cheap, and thus widely distributed, newspapers. Mass readership was made possible through the invention of the cylinder and rotary presses by mid-century, the invention of the telegraph, the reduction in paper prices, and the invention of photography during the second half of the century. Newspapers were able to introduce a multispatial present in word and picture that was both complex and explanation-begging. Time conception thus became "progressive," coupled with a monochronic use of time by individuals; in other words, people tended to be engaged in only one activity at any time. As such, societal linear and individuals' monochronic times became structured, and event order was now more inelastic.[29]

The changes described so far involved two major aspects. First, since technology dictated temporality modes to a large degree, it was turned into "applied time control"; second, productivity became the measurement and the organizing principle for both public and private goals. Thus, society moved from a "time surplus" to a "time affluent" stage. Time was related to as a resource, the more efficient use of which could lead to increased productivity. As a result, time became a major dimension around which society was organized.[30]

Twentieth century society responded to the transformation of temporality with three major processes (which could, of course, be related to other factors as well): first, a continued trend of industrial-technological innovations and modifications; second, the introduction of automobiles, especially the private car, as a more rapid and more flexible mode of transportation; third, a sophistication in the division of labor that called for more separation in time.[31] In Mumford's view, "only second in importance to the discovery and utilization of electricity was the improvement that took place in the steam engine and the internal combustion engine -- Neotechnic transportation awaited this new form of power."[32] The technological development and sophistication of the car was extremely fast, and was coupled with an intensive diffusion into households. Thus, maximum car speed in 1894 was 12 m.p.h.; in 1899, 39 m.p.h.; and just four years later, the 60 h.p. Mercedes was capable of speeds of up to 80 m.p.h.[33] In 1910, Ford started to market the model T. By 1927, 55.7% of all American families owned a car, and in 1930, this percentage increased to 87%.[34] U.S. President Warren Harding had told the Congress back in 1921 that "the motorcar has become an indispensible instrument in our political, social and industrial life."[35] In Europe, developments did not proceed at the same pace, and the ideal of a family car as a value of mass culture had to await the post World War II era. Thus, in 1930, there was one car for every 42 inhabitants in Britain and one for every 135 people in Germany, whereas the figure was one for every five people in the U.S.A. By 1970, the ratios had been closer to one another, the figure being one car for every five, four, and two persons in Britain, Germany, and the U.S., respectively.[36]

Concomitant with the rapid adoption of the car as a popular transportation vehicle by the turn of the century came a diffusion of the telephone as a popular means of communication. In Germany, for example, there were 71,000 telephones in 1891, the number growing to 1.3 million by the eve of World War I. By that time, there were already 10 million telephones in the U.S.[37] Continued pressures for higher productivity, for further division of labor, and for time-saving technological innovations made time an ever more important resource. An American consequence, in terms of a business-time evaluation, was Benjamin Franklin's "time is money." Time could, thus, be saved, sold, and shortened.[38] Society became time hungry rather than "time affluent," with a constant scarcity of time.[39] This ever growing demand for further time-saving has surely assisted the telecommunications revolution, which had begun in the sixties. Telecommunications technologies consist of improved and reasonably priced long-distance telephone service (even via satellites) and modern computers integrated into the communications industry, especially through two-way cable TV.[40] These new tools, rather than easing "temporal tensions," may cause additional demands on or for time, since many transactional activities in both economic and social life can now be performed instantaneously. Modern electronic communications has influenced the social significance of the present in terms of its speed, form, and distance.[41] Social demands for more time yielded the "colonization of time," making cities function 24 hours a day.[42] In addition, "time density," or "time intensity", has increased, so that efficient time use has become a constant societal goal.[43]

Since the turn of the nineteenth century, two types of times have evolved, public and private. Public time has to do with urbanization, industrialization, and transportation and is homogeneous. Private time, on the other hand, is heterogeneous, fluid, and reversible. Typical of both times was their having a strong future orientation.[44] Society at large, however, has more time in comparison with the limited time availability of individuals; the durational existence of the former extends beyond that of the individuals of which it consists. Time planning in various forms has penetrated into the life of both individuals and society. Individuals are engaged in the planning of daily and weekly calendars as well as their studies and careers. Society engages in production planning (in economic firms) and social planning (in social and political organizations). This planning is accompanied by a constant two-way push between individuals and society in capitalist cities for a more efficient time-use. The limited availability of time as a production resource for individuals, on the one hand, and the constant capitalist desire to see gains and profits after a short period of time, on the other, have made society (as a nation or a city) act as if it has a limited time-span, as well.

A further implication has to do with the question of time-use for leisure and relaxation. One could argue for the polychronic use of time for leisure purposes, even in modern societies, in the form of "simultaneous consumption," such as parties, where eating and meeting other people are performed at the same time.[45] But one may ask whether temporality in these activities really is very extensive and dispersed, or whether it is because we like "to get the most out of it" that leisure temporality is intensified in some instances.[46] An interesting analysis of leisure versus work times was provided by Diesing, who identified three stages in the relationship. In traditional societies, no clear distinction is made between leisure and work times. In an intermediate stage of modern societies, the two times corresopnd to two phases of the economy, so that the value of work time is known but the money-worth of leisure is not. In the third stage, the temporality of leisure becomes more rationalized and the distinction between work and leisure times gradually disappears; the result is that "in a pure, ideal economy there would be only one kind of time and one kind of calculation to apply to all activities."[47] From a social viewpoint, some people may tend to extend their work time into leisure time without clear boundaries, between the two, while others may relate to the two times in opposite or neutral modes. People engaged in research and development activities are examples of the first, while most production workers belong to the second group.[48]

At this point, it is interesting to mention Toffler's observations on future temporality, which should be affected mainly by the Third Wave, namely, the computer-telecommunications revolution (following the agricultural and industrial waves). Toffler claims that temporality may be changed, first, because of the relative independence of workers using these new technologies and the lack of any need for production synchronization. The importance of punctuality, for example, may thus be reduced. Second, Einstein's notions of relative time may bring about some yet-unknown changes in societal temporality, similar to changes that occurred after Newtonian absolute time was introduced (flexitime for work being an example of temporality changes related to relative time[49]).

On the cultural level, it has been argued that Einstein's ideas contributed to the creation of a synchronous time-space in the arts and mass-media.[50]

In summary, then, time has increasingly turned into a most important dimension and resource for modern urban society. Its conception along with its intensifying use, has become mostly linear. The bottom line of the transformations presented so far is that several social-spatial outcomes exist. Some of these, which are beyond the scope of this volume, relate to social stress. To mention just a couple of related aspects, there is the problem of individuals who have to adjust to increased pressures on their social time, which is channeled through intensified societal time-use. In addition, there is the problem of minority groups within a given society that function under different societal times, and the discrimination against such groups through an unequal intensification of time in the provision of welfare services. Giddens referred to this problem in his identification of front/back regions in the city.[51] Still another outcome, one that constitutes the focus of this chapter, is the possible changes in spatiality that may both result and influence the changes in temporality. Before going into this time-space comparison, let us first take a look at spatiality changes typical of modern urban society.

From Space to Space

Opening standard texts in urban geography, especially those with a North American accent, one would find the traditional discussions relating urban spatial growth and the deconcentration of population and economic activities, on the one hand, to developments in transportation technology, on the other. Standing at the center of such discussions is a spatial entity, namely the city, a major growth factor of which is transportation. This "spatial view" has been expressed in its extreme in the statement that "the friction of distance is the fundamental axiom of all geographical theory."[52] The approach to urban space taken in this book will, rather, attempt to discuss changes in societal spatiality in capitalist urban societies that have undergone socio-economic and transportation transitions. The focus will thus be on urban society; and transportation and communications will be viewed as enabling means for the achievement of societal goals and ideals. In the long run, a constant two-way relationship evolves between industrialization and urbanization, on the one hand, and the development of transportation and communication technologies, on the other; each pushes for further development in the other.[53]

Generally speaking, the change from agricultural and town economies to industrial and post-industrial involved a transformation from an intensive-space to an extensive-space society. Two things are meant by this assertion. First, the town urban society used quantitatively less urban space but relatively more intensively than do modern societies. Secondly, space has, qualitatively, a deeper, more varied significance for traditional than for current urban societies. These changes are outlined in Figure 4 and elaborated upon in the following paragraphs.

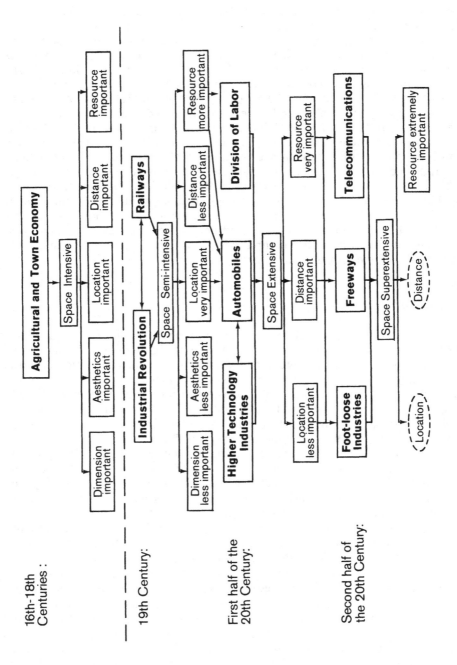

Figure 4:   Space in Capitalist Urban Societies

Until the nineteenth century space presented at least five meanings for urban societies, which as a result were heavily dependent on space. First, space was a major dimension around which everything else was organized. "One's land is the home of the ancestors; a heritage to be passed on to one's descendants. The unity of the place symbolizes the infinite unity of the society."[54] This significance of space at the national level has been termed the "powerful organizing metaphor."[55] Second, urban space was a source of aesthetics and cultural values, either in its natural topography of hills and rivers or in its built landscapes, which consisted of a variety of street systems, stairs, and major landmarks like cathedrals.[56] Third, locations within the city, or absolute locations, were of extreme importance for residences, businesses, and manufacturing plants. The strong dependence on limited transportation and energy sources made location within the city a crucial element in urban life. Thus, manufacturing tended to be located at the waterfront and the rich preferred central residential locations. Fourth, distance has been a major consideration in almost any daily social or economic activity, determining as it does relative location within and outside cities. In 1815, only about one person in fifty commuted as much as one mile to work.[57] It is not surprising, therefore, that early location theory (e.g., Von Thunen), emphasizing the importance of sites and distances, evolved under circumstances such as these. Finally, space as a resource. Outside towns, it was a productive, active resource for agriculture; inside towns, a more "passive" resource. Every piece of land had a value determined by its location and other factors.

The industrial and railway revolutions in the late eighteenth century began to change the attitude toward space in cities. The major transitions in the availability of mobility modes and energy sources increased the spatial horizons of urbanites both within and outside cities. Overcoming spatial friction became easier, while the time resources needed for the new spatial mobility increased in importance. The most important outcome was a tremendous urban growth in general, and in major cities in particular. By the end of the eighteenth century, only one sixth of the British population was urban, compared with three quarters a century later. In France, the respective figures were one seventh and two fifths. During the first half of the nineteenth century, the population of London trebled, and that of Paris doubled. By 1891, there were four million people living in London and 2.5 million in Paris.[58] Urban areas have not increased at the same rate, however. Population density in central London increased 2.5 times during the first half of the nineteenth century. In central Paris, population density decreased during the first half of the nineteenth century but increased during the second half, when railroads were introduced.[59] The importance of space as a major organizing factor or dimension thus decreased, though not yet in a decisive form, since urban densities were still high because of transportation limitations. At the same time, however, the importance of time increased, as we have already seen.

Productivity and efficiency as social values replaced urban aesthetics, during the late nineteenth century so that, especially in North America, very condensed and polluted downtowns developed, hills were removed, and standardized grid street systems imposed.[60] This process was augmented with a flood of industrial products, which caused functional architecture to develop. Europeans soon lost their "dignity of space."[61] One could interpret these changes as a linearization of space paralleling to the linearization of time.[62] The importance of location, meanwhile, increased, since railways tend to serve areas rather than points. The value of urban areas next to railway stations increased for both residence and production. Distance, however, decreased in importance, since the new technology permitted one to go farther than before. Distance was now often measured in temporal rather than spatial units. Urban spatial growth and the demand for more land increased the value of space as a resource, becoming what has been called "created space" and replacing the earlier "effective space."[63] All in all, nineteenth century, railway-based urban society could be termed as semi-intensive space.

The urban society in the first part of the twentieth century could be termed an extensive space because of the suburbanization process that typified it.[64] This process became possible with the introduction of the automobile and electric power. The first permitted a high level of spatial flexibility, serving points rather than areas, while the second reduced the locational dependence of manufacturing in the city. One may argue that these innovations were successfully adopted, among other reasons, because of social and cultural needs for spatial flexibility. Generally, however, space has not been completely replaced by time as an organizing metaphor,[65] but urban land and building aesthetics have been replaced by economic efficiency and the financial value of land. At the same time, increased mobility reduced the importance of the location and distance aspects of space. The importance of space as a resource, however, increased greatly. Not only did further urban spatial growth and the suburbanization process require more space, but the heavy use of the automobile required much more land. For example, it has been estimated that about 60-70 percent of the area of Los Angeles is devoted to cars (roads, freeways, parking lots, etc.), leading one writer to call the car the largest space consumer ever created by man.[66] The territorial size of Los Angeles vastly increased from 85 square miles in 1910 to 440 in 1930.[67]

Further technological innovations in the second half of the twentieth century turned urban society into a super-extensive space. This society enjoyed the use of innovations -- such as urban freeways, foot-loose industries, and telecommunication devices integrating telephone, computer, and television networks -- that clearly responded to the need to overcome location and distance obstacles. It is quite obvious that in today's information-based urban society, which is almost free of location and distance considerations, the use of space has received a completely different magnitude. Urban fields, daily urban systems, and exurbia have replaced the metropolis and its suburbs as the spatially organizing framework of urban areas. Space in this society may still be of extreme importance, but only as a resource. Webber noted the peculiar nature of land "characterized by increasing supply and ever-declining value" because of urban growth. Harvey termed this neo-classical view "an innocent trap," since Marx showed the limited supply of agricultural land.[68] It is important, however, to recognize some differences between rural and urban land. Rural land is restricted in terms of its quality for cultivation, availability of water, and distances to the farmer's house and to the market. Urban land, which is "passive," is restricted mainly by accessibility. Assuming the use of modern transportation and communication technologies, accessibility is reduced in importance; it is planning that now becomes the major limiting factor for further urban expansion.

The extremely extensive use of space by neo-capitalism has been viewed as an "urban revolution" equivalent to the industrial, with space being considered a social product or an output rather than a resource or an input.[69] These two options should not be interpreted to be contradictory, but rather complementary. Space must be assessed by society as an important, crucial resource before it may turn into a product. Urban life calls for a constant process in which space is changed from a resource into a product through land-uses and the social significances attached to them. Market and planning regulations serve as regulating mechanisms in this transitional process from input to output.

In summary, space gradually lost its importance for urban society in several respects; however, its value as a resource increased, along with a more extensive use. The transitions in spatiality outlined above involve several social and spatial results. On the micro social side, which again remains beyond the scope of this volume, impersonification of human relations is the most obvious. Face-to-face contacts are reduced when cars and telecommunications are used for business, shopping and entertainment.[70] At the spatial end, the very inefficient use of space in urban areas seems obvious. Although it is possible to argue for a potential difference in time-use between business and pleasure, there is no difference in the wasteful use of space for the two functions. Hence, second homes and weekend resort areas may be located even beyond exurbia.[71]

Temporality Versus Spatiality

By comparing Figures 3 and 4, the reader can guess where this section is heading. A comparison between time and space may be made concerning the changing relationships between both temporality/spatiality and temporalization/spatialization. First, though, some comment has to be made regarding the relevance of time-space analysis at the urban rather than at the national level of modern capitalist societies. It has been observed that "condensation and storage, for the purpose of enlarging the boundaries of the community in time and space, is one of the singular functions performed by the city."[72] On the other hand, the argument is also advanced that the nation-state replaced the city as "the dominant time-space container."[73] Nevertheless, most human and societal activities are urban in nature, and most of the activities of nation-state institutions take place in cities even though their perspectives might be national (or international).

Figures 3 and 4 obviously present two inversely related change-processes in temporality and spatiality. Whereas temporal efficiency, density, and intensity have increased, these respective spatial characteristics have decreased. Moreover, these immense trends evolved together with similar social and technological changes. In the early stages of capitalism, the two resources, time and space, received partially similar treatment, so that when time-use became intensified, urban areas suffered from increased density (which was even more amplified with the introduction of the elevator in the late nineteenth century). Even then, however, an extensive use of urban space started to emerge through the construction of suburbs. In late capitalism, time-use has become extremely intensive, and space-use very extensive. If time and space had been treated by society in the same way "time-wise," then we would have witnessed, in advanced capitalism, extremely dense cities containing only high-rise buildings. Why is it, then, that in early capitalism and, much more so, in the twentieth century, spatiality became extremely extensive while temporality became more and more intensive in nature? And why does it seem as though these two processes are interrelated? These questions are posed mainly with regard to space, since we would have expected that dimension to behave in a fashion typical of classical capitalism, which calls for an intensive use of resources.

The explanation lies in a blend of two basic social elements: culture and economics. One may recognize that European metropolitan areas have exhibited spatial suburban growth trends later than did North American cities.[74] These trends, however, do not only reflect differences in national areal size, they are also expressions of cultural values and societal norms. Early in their history, Americans were noted to be characterized by a love of newness, a desire to be close to nature, a sense of freedom to move and by individualism. These values are still expressed in current American landscapes.[75] It was, however, through social pressures that the "American dream," centering on a dream house, ignored alternative hopes for ideal cities or for an efficient consumption of scarce resources.[76] The rise of privatism in American society especially emerged in the middle third of the nineteenth century and was associated with wealth. This phenomenon further reflected the increasing importance of family life (which will be delved into in more detail in the next chapter). It was the private yard around the suburban house that provided for a form of individualism, or separateness.[77] Although the British shared this need for privatism, as expressed in the planning tradition of country-town integration, Britain's long planning history was not imposed on North America, thus permitting less-controlled spatial growth in the New World.[78] Recent European suburbanization and exurbanization might be interpreted, therefore, as a transatlantic diffusion of the "American dream," brought about by American dominance in modern mass-culture, especially via television and the movies.

No less important to the extensive mode of urban spatial growth is the expansionist nature of capitalism. Expansion into time has been limited, as mentioned previously, because of the willingness to produce a great deal and fast and because of the expectation of receiving profits and gains as soon as possible. The expansionism into space could take a different, less-efficient, form, since the ownership of more land means the control of more resources.[79] Moreover, land ownership became a mark of status. This has been a European value since the Middle Ages and it was carried over into the New World and into capitalism.[80] An intensified use of land could, though, offset this trend and decrease the quality of life. Thus, capitalistic norms for individual temporality and spatiality made society at large adopt temporality and spatiality norms, which do not fit the realities of societal time and space inventories. Time has been used as though there is not enough time available, although in absolute terms society has infinite amounts of time. Space has been used as if it is unlimited, which it is not. It only seemed that "the real estate of North America was almost literally endless."[81] The same argument holds for temporality and spatiality with respect to both production and consumption. For example, in industry, production may be spread over vast areas in metropolitan cities, thus representing an extensive spatiality. At the same time, research and development are speeded up even if there is no real social need for a new product or service, thus showing intensive temporality. By the same token residential values may call for the further spread of cities even if space is limited, demonstrating extensive spatiality. On the other hand, shopping and the electronic media function 24 hours a day, exemplifying intensive temporality.

The time-space inverse relationship described here is probably true only for urban industrial and post-industrial societies. In agricultural societies, it has been noted, "electric light and irrigation as technical innovations have two interesting features in common. Both allow temporal expansion in resource occupation, but both also result in a parallel contraction in spatial distribution and accessibility. Temporal expansion is constrained to those spatial areas which are covered by the irrigation network and electricity network."[82] Thus, agricultural societies present an inverse time-space relationship, in which time-use is extensive and space-use intensive. One should note, though, that the preceding comparison involved the spatial and temporal outcomes of two innovations connected with the use of resources in agricultural societies. Our argument is that the spatial and temporal behavior of urban societies is interrelated because of technological changes. Attention has also been given to the synchronization (in time) and synchorization (in space) of inputs such as energy and water in agricultural societies.[83] In urban societies, however, time-space changes follow sequentially changes in communications technologies; therefore, a more extensive use of space may be permitted, followed by a more intensive use of time.

The contradictory inverse time-space relationships between urban and rural societies may stem from the different uses of land and time in the two societies. In agricultural society land is a productive, "active" factor; intensification means more yield from a given unit of land. In urban society, land is merely a locational, "passive" resource for all urban components. Time in agricultural economies is also different from time in urban industrial and transactional economies. In the first, time is measured in seasons and sometimes in years (for plantations); whereas in industrial production, time is measured in hours, and in transactional processes it is rather instantaneous.

Transportation and Telecommunications

We have seen that the two inversely related processes of temporality and spatiality are interrelated through similar developments and uses of transportation and telecommunications. Innovation in these areas has influenced both temporality and spatiality, though in opposite ways. As time usage becomes more efficient, more pieces of space may be "overcome" in terms of distance and used in terms of resources and locations. When more land is used, and more extensively so, time usage must become even more efficient in order to facilitate convenient accessibility to and from the new areas. This vicious circle has been moving faster and faster with further developments in transportation and communications, and hence the extreme pressure for still further innovations and developments.

Janelle's famous "time-space convergence" model for the urban scale deserves a second look, using the observations made here.[84] Assuming an intensifying temporality coupled with spatiality becoming more extensive, could one then assume the existence of an urban "time-space convergence"? The meaning of space might provide an answer. The "time-space convergence" model interprets space as distance, so that with time intensification (via transportation improvements), there is "space (distance) shrinking." This view however ignores, the more extensive use of urban space as a resource, which has been noted here as an outcome of time (as a resource) intensification.

Temporalization and Spatialization in Urban Society

The spatialization of time in capitalist societies, as mentioned earlier, is a process propsoed by Gross, who views this spatialization as a weakening of the temporal dimension.[85] Before turning to a detailed discussion of this view, a more basic question has to be readderssed; namely, the relative importance of time and space in capitalist urban societies. One writer discussing only space, refers to space as the "seat of power."[86] Others, discussing both time and space, see time as replacing space as the more important dimension: "The spatial milieu of our activities becomes less important than the social or temporal milieu."[87] In this chapter, we have attempted to show the increased importance of time and the interrelated decreased importance of space. This change probably developed once time started to be valued as resource, in the nineteenth century. Moreover, one could argue, it was then that urban society began to replace "spatial tyranny" with a "temporal tyranny."[88] From a dependence and appreciation of space for versatile city development and livelihood, society turned to time-dependence in almost all aspects of urban life. Time became a resource, a container, and an important dimension.

The spatialization of time, it was argued at the beginning of this chapter, was coupled with a temporalization of space. Gross saw this spatialization process as the use of spatial modes for time conception and measurements.[89] As such, it was considered a negative process. As we have seen, time has become more important than space as a leading dimension and resource in modern urban life. The view of spatialization should be broadened, therefore, to mean that time has taken over the importance formerly given to space. The spatialization of time also means that time is used intensively the way space was used in the past. Thus, today's temporality is imbued with both the form and the importance of spatiality of the past. In this light, spatialization of time carries with it not only a negative connotation but a positive one, as well (from a temporal viewpoint, of course).

The temporalization of space means that space is now measured, and sometimes even conceived, in temporal modes. Thus, distances, especially within urban areas, are measured in time units rather than in distance units. Moreover, as we have seen, urban space is conceived of and used the way time was used in the past -- namely, in an extensive mode. In other words, current spatiality has the form of early temporality. Temporality and spatiality of capitalist urban societies, therefore, have undergone a full-cycle exchange of modes, so that time has been spatialized and space temporalized.

Time-Space, Planning, and Technology

A hidden assumption lies behind the analysis presented so far, it is that society uses time and space in free and uncontrolled ways. The trends presented here are true for such capitalist urban societies in which societal as well as ethnic or religious controls over further urban expansion are low. The powers and means to control time-use and time-intensification are not so well understood; consequently, "we have land use planning but seldom time use planning."[90] The capitalist norm for an efficient time-use does not have general rules and regulations administered through formal institutions. Specific production standards do not apply beyond a certain industry or after working hours. The controlling mechanism of space-use is urban and regional planning. The temporal equivalent of spatial planning might be five-year plans, which are common in socialist regimes.

Time and space controlling mechanisms raise some interesting questions. What happens in a non-socialist society with a strong planning tradition, such as Great Britain? Should we expect less pressure for the adoption or development of time-saving communication technologies because of societal limitations on the possibly resulting urban growth? If so, may one then speak of spatial planning as an indirect form of temporal and technology planning? By the same token, what might the recent societal intervention in spatial growth in North American cities through governmental and agricultural land protection do to time and technology use? Is the "hunger" for further spatial expansion just a side effect of time saving, so that further time intensification and higher productivity are independent of space? Or is there a two-way relationship?

As already noted by students of industrial capitalism, industrial innovation is related, to one degree or another, to the capitalist system insofar as capital accumulation calls for the production of new products, thus causing a faster pace of the economy.[91] The process of innovation, therefore, is at least partially independent of temporality and spatiality. One could, though, still hold to the view that intensified temporality is both a cause and an outcome of industrial innovation. This is due to the role of time-use as a key factor in capitalist production, which we have pointed out.

The control of space has potentially little impact on time control although the spatial construction of a new road, for example, causes time savings. The relative lack of impact of space control on time control stems from intensified temporality being mainly the result of capitalistic norms for faster production and consumption. Thus, time-use is not necessarily space-dependent. Time control has, however, a potential impact on space control. If time control would have been possible, space-use would have been affected directly, since it is time-saving that permits a more extensive use of space, and vice versa. An example of existing controls are driving speed limits. These limits (or the unavailability of an expressway) may obviate or limit the use of outlying areas in an urban context. Time-control mechanisms, could have offered another aspect of the spatialization of time, since time-planning would then have been the equivalent of urban spatial planning. On the other hand, the current existence of space-control and the almost total lack of time-control at the macro-level of capitalist urban societies comprise another aspect of the inferiority of space vis-a-vis time in such societies.

Conclusion

The preceding sections have shown a structuration process in which spatiality and temporality have undergone major changes simultaneously with societal, economic, and technological developments. These societal and technological developments were mutually reinforced with changes in temporality and spatiality. The structurationist process of "expansion" into time and space accompanied the evolution of capitalist and technologically advanced modern urban society. Thus structural and technological developments caused new modes of temporality and spatiality. These new modes, in turn, called for further structural and technological changes. It is impossible, therefore, to separate social structural change from time and space, and vice versa. The societal "expansion" into time and space and the new modes of spatiality and temporality led to a new relationship between time and space. Temporality has intensified, while spatiality became more extensive. The inverse relationship between the two seems to be a one-way phenomenon, since an increase in spatial density might not necessarily cause a decrease in temporal intensity. On the other hand, an intensified use of time permits and calls for further use of space, since more distance/space may be "overcome" in a given time unit when time-saving technologies are introduced. This argument appears to contradict the common phrase "time-space", which implicitly assumes a similar behavior of both aspects. Although the "time-space" phrase may fit a behavioral analysis of individual urbanites, an examination of time and space of urban societies at large requires a careful distinction between the two.

Besides the contradiction between time and space at the urban societal level, there is also some complementarity or convergence between the two. Durkheim suggested that "the concept of time is reflected in the organization of space."[92] Following this hypothesis, the circular plaza of polychronic Southern Europe has been contrasted with the linear Main Street of North America. The former permits talking to several people at a time, the latter reflects a lack of interest in others.[93] One may conclude, therefore, that at the micro level of the built environment, there is a time-space convergence; at the macro level, there is a time-space divergence. Some interesting relationships also exist between the social-micro, (namely, individuals) and the social-macro, (namely society). One such may be discoverd when time and space are studied along gender. The time and space of men and women constitute the topic of our next chapter.

NOTES TO CHAPTER 3

1. See Urry, 1981.
2. Pred, 1983; 1984
3. Kern, 1983, p. 4.
4. Braudel, 1980; Gross, 1985.
5. Discussions on the transformation in attitude to time are presented by Thompson, 1967; Thrift, 1981; Giddens, 1979; 1981; Whitrow, 1972; Zerubavel, 1981; Toffler, 1980; Lowe, 1982; Kern, 1983; Landes, 1983. The topic was also mentioned by Linder, 1970; Ilchman, 1970; Hall, 1966; Forer, 1978; Kolaja, 1969; Carlstein, 1978; Mumford, 1934; Innis, 1951; Heller, 1978; Zentner, 1966; Diesing, 1962.
6. Ilchman, 1970.
7. Eliade, 1954.
8. Hall's observation that polychronic time is rather casual and unstructured fits the social and economic scenes. It is questionable, however, whether it would be true of religious rites, as well.
9. Parkes and Thrift, 1980, p. 87.
10. Linder, 1970.
11. Whitrow, 1972, p. 7.
12. De Grazia, 1962.
13. Giddens, 1979, p. 201.
14. Whitrow, op. cit., p. 6.
15. Giddens, 1981, p. 133.
16. Landes, 1983, pp. 60; 69.
17. Whitrow, op. cit.; Giddens, op. cit.
18. On the rising interest in astronomy see Mumford, 1934, and Whitrow, op. cit. Puritanist morals and time are discussed by Thompson, 1967, and cultural-artistic developments may be found in Heller, 1978. The invention of the mechanical clock is detailed in Mumford, 1934; Thrift, 1981; Landes, 1983. Christian traditions and time were mentioned by Giddens, 1981, and by Landes, 1983. The consequent faster time-space is elaborated by Mumford, 1934.
19. Landes, op. cit., pp. 89; 228; 230.
20. Giddens, 1981.
21. Thompson, 1967; Thrift, 1981.
22. Giddens, 1979; p. 201; 1981, pp. 130-131.
23. Zerubavel, 1981, pp. 59-64; Toffler, 1980, pp. 64-67; 115-117.
24. Lowe, 1982, p. 11.
25. Kern, 1983, p. 111.
26. Mumford, 1934, p. 199.
27. Landes, 1983, pp. 93-94; 285-287.
28. Lowe, 1982, pp. 36-37.
29. On newspapers, time and space see Lowe, 1982. The "progressive" time conception was propsoed by Ilchman, 1970, and the monochronic use of

time was put forward by Hall, 1966. On the inelasticity of event order see Parkes and Thrift, 1980.

30. Ilchman, op. cit., p. 138; Linder, 1970.
31. On division of labor and separation in time see Goody, 1968, p. 39.
32. Mumford, 1934, pp. 235-236.
33. Pettifer and Turner, 1984, p. 38.
34. Flink, 1975, p. 142; Schwartz-Cowan, 1983, p. 83.
35. Flink, op. cit., p. 140.
36. Jackson, 1985, p. 163.
37. Kern, 1983, p. 214.
38. Soule, 1955, pp. 90-93.
39. Linder, 1970; Zerubavel, 1979.
40. Kellerman, 1984.
41. Kern, 1983, p. 6.
42. Melbin, 1978.
43. Forer, 1978; Kolaja, 1969; Carlstein, 1978.
44. Kern, 1983, pp. 34-35; 94-95.
45. Linder, 1970, pp. 77-109; Zerubavel, 1981, pp. 58-59.
46. Hall (1966) would argue for an extensive and dispersed time-use in leisure activities, while de Grazia (1962) and Bell (1973, pp. 472-475) would argue for an intensive production-like temporality even in leisure activities.
47. Diesing, 1962, p. 28.
48. Parker and Smith, 1976.
49. Toffler, 1980; pp. 264; 306-309.
50. Lowe, 1982, p. 6.
51. Schwartz, 1978; Giddens, 1979, p. 209.
52. Falk and Abler, 1980, p. 62.
53. Kellerman, 1984; Lowe, 1982, p. 37.
54. Diesing, 1962, p. 29.
55. Ilchman, 1962, p. 138.
56. Katznelson, 1979, pp. 230-231; Mumford, 1934, p. 18.
57. Jackson, 1985, p. 15.
58. Lowe, 1982, p. 36.
59. Clark, 1951.
60. Katznelson, op. cit., pp. 230-231; Jackson, 1985, p. 74.
61. Kern, 1983, p. 156.
62. Toffler, 1980, pp. 120-121.
63. On "effective space" and "creative space" see Harvey (1973).
64. Toffler, op. cit., p. 309.
65. Ilchman (1970). Different observations were made by Lakoff and Johnson (1980).
66. Hall, 1966, p. 175.
67. Jackson, 1985, p. 139.
68. Webber, 1963, p. 42; Harvey, 1973, p. 178.
69. Lefebvre, 1970; 1976; 1977.
70. Hall, 1966, p. 75.
71. Lefebvre, 1970, p. 265.

72. Mumford, 1961, p. 98.
73. Giddens, 1981, p. 147.
74. Hall and Hay, 1980; Van den Berg, et al., 1982.
75. Crevecoeur, 1782. On current expressions of these values see Meinig, 1979.
76. Hayden, 1984, p. 38.
77. Jackson, 1985, pp. 58-59; 78-79.
78. On the British planning tradition and privatism see Lowe, 1982, pp. 68-69.
79. See also Lefebvre, 1970; 1976.
80. "The American dream was in large part land" (Jackson, 1985, pp. 53-54).
81. ibid., p. 129.
82. Carlstein, 1978, p. 156. In another place, however, Carlstein (1982, p. 259) claimed that irrigation caused spatial expansion, as well. Giddens (1981, p. 94) noted, though in passing, the "greater coordination of time-space relations" in irrigation agriculture.
83. Carlstein, 1978, p. 156.
84. Janelle, 1969.
85. Gross, 1981; 1982. This concept was also mentioned in passing by Soja (1979, p. 3). Lowe (1982, pp. 42-43) discussed the cultural and historiographic despatialization of time in bourgeois society.
86. Lefebvre, 1976, p. 83.
87. Wise, 1971, p., 21. See also Ilchman (1970, p. 138) on the national scene, and Lowe's (1982, p. 11) cultural analysis.
88. The term "spatial tyranny" was proposed by Warntz, 1968.
89. Gross, 1981; 1982.
90. Parkes and Thrift, 1980, pp. 101; 112.
91. E.g., Mumford, 1934, pp. 28-29; Webber, 1982.
92. Durkheim, quoted by Parkes and Thrift, 1980, p. 87.
93. Hall, 1966, pp. 173-174.

CHAPTER 4

TIME AND SPACE: MAN AND WOMAN

We saw in the preceding chapter that societal temporailty and spatiality have undergone extreme changes under capitalism. Temporality changed from extensive and cyclical to intensive and linear, while spatiality was transformed from being intensive around small spaces to being extensive around large areas. The purpose of this chapter is to examine these trends along a gender differentiation between men and women. The major arguments to be presented here are that men and women do not necessarily share the same temporality and spatiality; also, that societal transformation of these basic aspects of humanity bear far-reaching consequences with regard to female needs.

An examination of spatiality and temporality along gender provides, more than does any other differentiation, a bridge between the study of the micro level, of individuals and the macro-societal level. There are at least two reasons for this. First, the gender differentiation of the human race is the most basic division, and consists of two equal-sized groups. Secondly, the study of temporality and spatiality along gender involves an examination both of personal-internal elements and experiences, though of a collective gender nature, and of external societal elements acting differently on men and women.

Geographers, in general, have not paid enough attention to women's geography. Indeed, it was only in the early 1980s that the first monographs on the geography of women were published. These texts, as well as several more specific studies, provide useful information on the spatial and temporal problems of women, some of which will be detailed later in this chapter.[1] What is still missing in the literature, however, is an interpretation of these problems in light of general societal transitions in spatiality and temporality.

We shall start with a short review of the evolving differences between men and women under capitalism, then follow with an analysis of the significance of these differences in light both of women's spatiality and temporality and of the transformations in societal spatiality and temporality. The implications of this analysis with regard to women and suburbanization will conclude the chapter.

Transitions in the Status of Females and the Family Status Under Capitalism

At least three avenues of explanation have been proposed in feminist and other social science literature for the disadvantaged economic and social conditions of women. One approach is the biological one: the claim that female inferiority is not of a social but of a biological nature, so that women have always been at the mercy of their biology, which assigned them the role of birth-giving.[2] A second approach relates women's condition to patriarchism, which has prevailed in both traditional and modern societies.[3] A third theory attributes their situation in modern societies to the transition from an agricultural-rural to an industrial-urban society of a capitalist nature.[4] Traditional society was characterized by women's integral participation in agricultural production, which was considered part of "nature." Women were technological inventors as much as homemakers. The latter responsibility was shared with grown children and members of the extended family.

Evidence from current traditional societies is less clear cut on women's status. Examples have been cited to show <u>varying</u> degrees of the <u>daily</u> participation of women in agricultural production. Some seasonal flexibility also characterized women's agricultural production, apparently, so that the division of labor between the sexes reflected not only norms and power but also "practical feasibility."[5] Differences between sex role patterns in traditional societies have been especially viewed along agricultural systems and social organization.[6] Thus, shifting subsistence cultivation was characterized by family manual labor so that women were put in the center of agricultural production. It is clear why Africa has been termed "the region of female farming <u>par excellence</u>."[7] Asia, on the other hand, presents a different pattern of female agricultural production, in which women work less or not at all in these activities. This pattern related to men's use of ploughs, the introduction of hired labor, and a more settled and less tribal social organization. By implication, the veils used by women in Asian societies are seen as an outcome of these economic differences, and <u>vice versa</u>.

The transition to industrial capitalism in the West or to cash cropping in the Third World has usually removed women from mechanized agriculture and caused an economic and social separation between public economic production and private family life.[8] Thus, capitalism separated production and consumption, turning the first into a complex pattern of interdependence among workers and leaving the latter, in its family form, at a relatively low level of external interdependency.[9] Whereas early, rapid, and massive industrialization and urbanization permitted an increased sphere of activity by females, later forms of urban growth (i.e., suburbanization) increased the geographical and functional separation between production and family life.[10] The recent growth in women's participation in the labor force has, thus, resulted in their double role as producers and homemakers, and in consequent role conflict.

The industrial revolution caused a sharp division of labor between men, who specialized in the production of products and commercially sold services, on the one hand, and women, who specialized in the creation and maintenance of lives and in related domestic services, on the other.[11] Traditional societies, with limited production and health technologies, have been more concerned with life in its bare survival connotation. Therefore, the naturally feminine role of life creation and the natural or social roles of homemaking have been more central to and interwoven with limited and more natural agricultural production. It is industrialization and its human-made, removed-from-nature, modes of production that have become masculine in character. From the perspective of post-industrial society, one may generalize that traditional economies have been feminine in nature, while industrial economies have been more masculine in character in their being removed from nature, and thus from the family. Post-industrial society is characterized by two aspects that might be interrelated.

One is the change of emphasis in production from heavy, semi-manual, industrial production to completely automated production and an economic concentration in services production -- trends permitting and requiring a larger female participation in the labor force.[12] The second is the women's movement which calls for social change with regard to the specialized division of labor instituted by capitalism.

As we shall see in the following sections, these major factors and the resulting transition in women's status are closely related to societal temporality and spatiality. A simple integration of traditional feminine temporality and spatiality with masculine capitalistic temporality and spatiality is, however, difficult to achieve.

Temporality and Women's Status

Cyclical time, which has been a basic characteristic of traditional societies, may be interpreted as feminine time. Although this statement may seem simplistic at a first glance, a look at women's lives and inner temporality may provide some interesting evidence for the assertion. Women more than men experience biological and interrelated psychological cycles in their lives.[13] These cycles are menstruation on a monthly basis, pregnancy and birth at different times and frequencies during their fertility age, and menopause at the end of this age. It might be argued that human lifetime as a whole may be considered one big cycle and that both men and women are subject to biological rhythms; nevertheless, a major undebated difference between the sexes is women's cyclical temporal aspects relating to their role in the reproduction system.

The cyclical societal time of traditional societies does not imply the adoption of human-female cycles for all life aspects, since agricultural seasons are not related to human female time. A society that is based on cyclical <u>natural</u> time however, agrees more with human-female time and sees it as an integral part of its life system. Menstrual periods were not hidden, therefore, but rather marked, and women's birth giving and infant-caretaking were integrated with their productive roles.[14] The temporal dimensions of life creation and maintenance, on the one hand, and those of economic produciton, on the other, adhered to the same frameworks of nature and cyclical time. Societal use of cyclical-female time in economic production did not automatically imply a high social status for women, since it was not economic production that yielded social prestige and decision-making in traditional societies. These latter rewards were allocated to the church, the courts, the army, and the government, and women played a less central role in them.[15] Ancient cultures were aware of the differences between the sexes with regard to temporality. Thus, traditional Judaism exempts women from religious obligations that have a temporal context. The Chinese-Japanese Yin/Yang dichotomies relate Yin to men and time, and Yang to women and space.

In contrast to cyclical time, societal linear time is less natural and rather human made. Regardless of the historical inception and origin of linear time, it is a result of <u>human</u> history, culture and technology (see Chapter 3). The change from a societal emphasis on cyclical temporality to a linear accent, mainly in production, accompanied the industrial revolution and the rise of capitalism. Linear time may thus be interpreted as masculine time. It mainly affected men, who even before the industrial revolution, were more involved than women in production. In industrial capitalism, economic production became more central in social life as a prestige-rewarding activity. Industrial production and capital accumulaton, however, do not subscribe to any preset cyclical natural order, and the so-called economic or business cycles are not natural. The organization of social and economic life along linear rather than cyclical time has made it difficult for women both to adjust to and attempt to integrate their cyclical times into a linear temporal environment. Moreover, linear time, by its very nature, may be intensified, which is exactly what has been done with time-use under capitalism. Cyclical time, on the other hand, cannot normally be intensified, since the natural cycles have a more or less fixed pace and duration. An increased gap has, therefore, developed between capitalist production life with its ever-intensifying temporal linear pace, and relatively fixed female cyclical temporality. The separation of production and consumption along sex and the different modes of time-use associated with each sex created the stereotypes of punctual men and always late women. Men have been used to strict time schedules, since modern production called for a strict temporal interdependence among workers; domestic consumption, by contrast, was more loose in schedule.[16] This last differentiation may reflect some basic differences between masculine and feminine inner temporalities; this topic deserves some elaboration.

Women's Inner Temporality

The discussion of temporality so far has concentrated on society at large. The two modes ot temporality -- cyclical and linear times -- have been discussed and interpreted as feminine and masculine times, respectively. A major social-historical trend, the industrial revolution, was shown to have caused a drastic change in societal temporality, with significant but different implications for male and female temporalities. What about the inner significance, perception, and experience of time by men and women? In other words, can one identify societal factors, such as capitalism, that are exogenous to individuals as determining temporality? Or are there differences in the inner-self between males and females regarding temporality? Can one identify any gender differences in temporality that are due to biological, physiological, and psychological factors? Unfortunately, the study of gender differences in temporality is surprisingly sparse; though, as we shall see later, this is not the case regarding spatiality differences. Perhaps this situation stems from the fact that "gender temporality" seems at first glance to be of mere academic interest in philosophy and psychology, while the study of spatiality has many applied implications for education, social work and geography, among others. Given our arguments on the gender implications of societal temporality and, later, on "gender spatiality," it might well be that a comparative look would raise some interesting questions.

An exception to the limited study of gender inner temporality is Cottle's extensive work on this topic.[17] Cottle's research documents gender differences in inner temporality through a study of American corpsmen and corpswomen. The possible factors for these differences, however, are presented as mere intelligent speculations requiring much more study.

Following Bergson, Cottle identifies two modes of time: Continuous, linear time, or temporality, is defined as "the continuous indivisible flow of time"; spatial time, or spatiality, is defined as "the conception of time as atomic and divisible."[18] Linear time is objective, spatial time subjective. (Obviously, "spatial" and "spatiality" here do not carry any geographical connotation.) In the linear conception of time, one cannot know the future; whereas in the spatial, atomistic conception of time pieces, we can correlate past, present and future moments, since this time mode is based on our involvement in these moments. The two modes of time should not be looked upon as contradictory, as spatial time is somewhat linked to linear time. The reason is that the distinct separation of past, present, and future moments does assume that they are ordered in linear temporal order. There are differences of opinion, however, on the question of which of the two modes takes precedence. Bergson, as interpreted by Cottle, argues for the absolute superiority of linear time. On the other hand, Kant's view was that the intuition of time is grounded in spatial terms. Cottle seems to side with Kant in this regard when he writes: "In large measure, we cannot experience the phenomenon of time until it is spatialized, until we 'detime' it."[19]

Interestingly enough, the distinction between linear and spatial time regarding an individual's time conception is similar to the social distinction between cyclical and linear time. In terms of time-mode, "linear" remains the same in each case, while cyclical time which recognizes specific chunks of time to be repetitive with similar events, is a sub-form of spatial time. It now remains to be seen whether spatial time is more typical of women, in parallel with our arguments on cyclical time, or whether both sexes live more by inner linear time. The latter option might be related either to basic human nature or to the increased emphasis on this time mode in modern capitalist societies. Surprisingly, the first alternative was found to be dominant in Cottle's findings. Thus, women tend more to perceive time spatially, whereas men have a higher tendency to perceive time lienarly. Women's spatial time is focused on the present. This same trend was also found in a study of the verbalizations of women in labor, which is a special time during which women tend to be self-concentrated.[20]

This general trend was shown by Cottle to consist of several aspects and characteristics. Both men and women would like to recover pieces of time periods from their personal pasts. Men wanted a year that they would have liked to rework or change. Women preferred just an hour or a day to be relived, but without its having a bearing on their present lives. Fear of change caused them to prefer small amounts of relived time. Women, according to Cottle, tend "to visualize their expectations as occurring in a series of interrupted presents and as ends in themselves, rather than as means towards some future-end which they create in the present.... For men, the present is a time of preparation; for women, it is a time of having."[21] As a result, he concludes, women do not fully believe that present plans may eventually affect future events. Paradoxically, the linear and flowing temporality attributed to men perceives the present as a "vital moment" linking the past with the future. Women would consider each of the three time zones separately. Women tend, furthermore, to extend the definition of present into the near past, thus disengaging themselves from the past, whereas men extend the present into the near future. Although women are more concerned with generational continuity than are men, women show less interest in the future. Women prefer the extended, thickened present. The masculine perception of the present is that it is instantaneous, depersonalized, bad, and weak - a truly linear conception of time.

When it comes to explanations of the different temporalities related to men and women, Cottle could offer only some good speculations. He tended to see the roots of these differences in socialized sex roles. Men are influenced by socio-economic matters, while women are influenced by social-emotional matters. Men are socialized into investment in the future through the development of a career; women have been socialized into maintenance of present conditions. Men, therefore, are taught to deal with the future; women learn to handle the present. Marriage tends to change both the present and the future for women more than it does for men. Women's temporality is much more sensitive to environmental temporality, or public time, which has to do with women's sharing. Men conceive of time as an economic resource, which is related to masculine exchange.

Only once did Cottle relate to an alternative explanation: a physiological-developmental cause. Women enter their childbearing age with the onset of menstruation, which is a clearly marked event. Men, on the other hand, develop a "fearful eventuality," since the beginning and end of their sexual potency are time-taking processes.[22] The question remains, therefore, whether sex-role differentiation in its temporal context is merely socialized or whether there is a social enhancing of some more basic biological and psychological aspects. Could one assume, for example, that women's greater concern with generational continuity has to do with their role in the reproduction system? By the same token, has the cyclical nature of women's reproductive system anything to do with their spatial temporality which consists of many extended presents? These questions do not intend to suggest that exogenous-environmental factors in socialized sex-role differentiation do not play an extremely important role in the shaping of male and female temporality. Rather, they are meant to demonstrate that the study of the causation of female temporality has still a long way to go.

A modern working woman must live and act under both forms of temporality (cyclical and linear times), while men act mostly along linear time. Production time changed from cyclical, natural, and extensive to linear and intensifying, whereas life-creation time remained cyclical. Bridging or integrating the two temporalities might prove difficult. But before turning to this aspect, the transitions in spatiality have to be identified.

Spatiality and Women's Status

Spatiality of traditional societies could be termed natural and feminine. It was natural, since, for the daily purposes of travel and service provision, transportation means were natural -- namely, walking and animal riding. As a result, villages and towns were small and concentrated, with agricultural fields located around the villages. Both work and services were located within easy reach of one's home. The small geographical units and the short travel distances permitted the integration of domestic housework, production, and service provision.[23] As such, traditional spatiality permitted women to be engaged in all spheres of life with little conflict.

This intensive, small scale spatiality was also life-oriented. Space or land was used for production purposes in a mainly active form: land produced agricultural products. The use of land for passive purposes -- that is, for the mere location of man-made objects -- was limited to housing and several public functions. Cyclical time left a strong imprint on the land through the seasons of year and their effect on agricultural production and rural life. Time, space, and economic life were thus integrated into one system that was of a more feminine nature and, as such, permitted the more natural participation of women in production.

As we have previously noted with regard to time, daily economic production in both villages and towns has not been the highest prestige-awarding activity. One could find extensive space-consuming activities, such as war-making, hunting, interurban, inter-regional and international trade, from which women were mostly excluded and which were considered of higher status. Thus, the use of small feminine spaces in daily life did not necessarily call for a high status of women; rather, it assisted women in their economic and domestic activities. By the same token, gender relations were frequently oppressive, despite the daily spatial integration of production and reproduction.[24]

The introduction of modern, human-made transportation means, the train and later on the car, made it possible for cities to grow beyond the natural walking and animal riding distances. In addition, the massive urbanization that went hand in hand with industrialization removed most workers from agriculture. Modern agriculture has also involved increased travel distances, with tractors, pick-up trucks, and jeeps replacing animals for transportation. These increased spatial horizons of both city and village are masculine in nature, since they mean a geographical separation of home and work. Furthermore, the high level of interdependence among production units and workers in both production and consumption, so typical of capitalist industrialism, has called for much movement in space. The lower level of external interdependence of domestic life in traditional societies did not require as much movement. A contradiction was thus created between the relatively small spaces of family homes of all kinds and the relatively large cities and agricultural areas. This spatial separation between production and domestic reproduction evolved gradually over the first half of the nineteenth century.[25] The latter "industry" remained feminine, the former became masculine. Because of their domestic responsibilities, women increased their spatiality much less than did men, resulting in shorter traveling distances and smaller action spaces for women.[26]

Modern family life means more residential space per family than in traditional societies. The spatial horizons of both production and consumption in the city, however, have increased to an even larger extent. Trying to bridge the smaller, feminine, domestic, local spatiality and the masculine, production-oriented, large spatiality may cause temporal and role conflicts. The longer time required for travel to work creates further pressures on female temporality, which has been shown to be strained in any case. Increased spatial separation between work and home may contribute to the role conflicts of the working mother and/or wife.[27] This internal conflict between the two forms of spatiality and its contribution to strains in temporality are female in nature and are shared by men only to a small degree.

Women's Inner Spatiality

The approach to female spatiality discussed so far has emphasized the domestic role of women, which is at least partially socially determined. Other approaches, however, attribute the more restricted female spatiality to women's lower spatial abilities. Spatial abilities consist of two distinct features; spatial visualization and spatial orientation. Spatial visualization is defined as

> an ability to mentally manipulate, rotate, twist, or invert pictorially presented visual stimuli. The underlying ability seems to involve a process of recognition, retention, and recall of a configuration in which there is movement along the internal parts of the configuration, or of an object manipulated in three-dimensional space, or the folding or unfolding of flat patterns. Spatial orientation invovles the comprehension of the arrangement of elements within a visual stimulus pattern, the aptitude for remaining unconfused by the changing orientations in which a configuration may be presented, and the ability to determine spatial relations in which the body orientation of the observer is an essential part of the problem.[28]

The study of gender differences in spatial abilities has focused mostly on spatial visualization ability.

Several writers have noted the poorer urban mental maps produced by females compared to males, and attributed this to less familiarity with the city because of less movement about it.[29] Others have found a similar type of difference between boys and girls and attributed it to differing environmental experiences and education.[30] A review conducted in 1974 of some 1,400 studies on gender differences concluded that males have a higher visual-spatial ability of a developmental nature, which reveals itself only after puberty.[31] These differences, it has been noted, while significant, are small.[32] The geographical implications for daily life of gender differences in visual-spatial ability have been rarely studied beyond the use of mental maps. An exception is Gilmartin and Patton's experiment involving the road-map reading ability of several dozen students. The researchers could not find significant differences between females and males.[33] Their findings, however, should be evaluated cautiously, since most of the participants were cartography and earth-science students, who might possess higher than average spatial abilities. (The small number of subjects in that study should wet the appetite of both geographers and psychologists to perform large scale experiments on the geographical significance of gender differences in visual-spatial skills).

When it comes to explanations of gender differences in spatial ability, there is no general agreement among social scientists on the factors and causes.[34] Geographers have not yet studied to any great extent the origins and implications of gender differences in mental maps and activity spaces.[35] At least four avenues of explanation, however, have been offered by psychologists: one psychoanalytic, two physiological, and one social.

Erikson concluded after observing children's games that "the girls emphasized inner and the boys outer space"; and "here sexual differences in the organization of a play space seem to parallel the morphology of genital differentiation itself."[36] Although this approach ignores the influences of socialization, it has served as a basis for a more general theory that holds that females attach less importance to issues that are beyond their reproductive system, which is considered "inner space."[37] A 1980 study seems to show that this is not necessarily so with regard to modern females.[38] In this study, however, the researcher, who tested adult females, referred to "outer" and "inner space" in their widest, non-spatial connotations. It might well be that in the strictest sense of "space", -- e.g., in children's games -- a difference between boys and girls can still be attributed to physiological factors.

Another theory posits that the spatial abilities of men are generally more developed than those of women because of differences in the hormonal system.[39] Women's spatial ability according to this theory has been found to vary with the menstrual cycle, being highest in the low estrogen phase,or the menstrual period. Estrogen more than testerone regulates or inhibits the monoamine-oxidase (MAO) enzyme level, which may cause changes in spatial ability. Since women have more estrogen, their spatial ability may be weaker than men's. Biologically, estrodial (E_2) is the most active estrogen, and this

is the biological agent mediating spatial ability. Cerebral E_2 is provided either directly from plasma E_2 or indirectly by conversion of plasma T. The central locus of E_2 is primarily old parts of the brain. The effect of E_2 on spatial ability is mediated through early adjustment of later central nervous system (CNS) steroid responsivity, and/or through early growth and organization of brain tissues essential for the processing of spatial information. There is an optimal range of cerebral E_2 values for the maximal expression of spatial ability. At puberty and perhaps also later in life, activational effects of E_2 may lead to permanent or temporary inhibition or facilitation of spatial ability, and, either too high or too low cerebral E_2 level minimize the expression of spatial ability.[40]

The effect of the optimal estrogen range (OER), in this theory,is to permit intra-gender differences in spatial ability; as a result of biochemical individuality expressed by levels of E_2, therefore, some women may be superior to men in their spatial abilities.

A third, neurological, explanation relates to brain structure.[41] The right brain hemisphere is said to dominate the behavior of males more than of females, and this hemisphere is responsible, among other things, for visuo-spatial tasks. There is also evidence that males have a stronger hemispheric specialization, which contributes to increased separate hemispheric functions.

The effect on spatiality of socialization in sex role differentiation has often been suggested to be most crucial, at least in the fostering of spatial skills.[42] The role of parents, especially mothers in determining more restricted spatial reaches for their daughters' games has been noticed. Boys' emphasis on spatial manipulations and girls' accent on interior spaces have been related to learned socialized sex roles. Interestingly enough, Saegert and Hart note that spatial abilities are transferred by mothers to sons and by fathers to daughters. Furthermore they report on spatial training that would reduce the initial inter-gender differences in spatiality.[43]

The social and biological-psychological origins of female spatiality might, however, be interrelated:

An innate tendency toward greater investigativeness and aggressiveness would lead most males to a higher level of environmental experience than that achieved by most females. For adaptive reasons, child training in most societies would support the innate tendencies and thereby amplify the original sex difference. An unintended and uninstitutionalized but widespread effect of this mutually reinforcing pattern would be the superior male performance found on spatial skills.[44]

Male Versus Female Temporality and Spatiality

Any list of male and female institutional metaphors relating to social attitudes, organization, modes of change, and other social aspects must also contain metaphors relating to time and space.[45] Males live by linear, ever-intensifying time, while women live by cyclical-spatial time. Females would prefer small, intensely used spaces, while men prefer larger, more extensive spaces. Males, it seems, have stronger spatial abilities. Women have a different time-orientation than do men in the sense that cyclical-spatial time is more meaningful in their lives than it is for men.

If, as some argue, female spatiality is tied to female temporality (through the menstrual cycle), then it should be important to see whether the opposite is true for males. In other words, is men's heavier emphasis on linear time influenced by their stronger spatial ability so that men's temporality is tied to their spatiality? Does a wider "spatial reach" influence a longer temporal one? If it is so, then may one conclude that a reverse temporality/spatiality relationship exists between males and females? If female spatiality is dependent to some degree on female temporality, would the opposite be true for men? The answers to these questions are obviously beyond the scope of this study and deserve separate attention.

The emphases of capitalism on linear time and extensive space have increased the inequality in economic opportunities open to men and women. As will be shown in the following section, it is difficult to close these gaps. Masculine-capitalist temporality and spatiality produced inverse relationships between an intensive temporality and an extensive spatiality. Those women who share men's time and space have to cope with these realities in addition to coping with the conflicts between men's and their own spatiality and temporality.

Spatiality, Temporality and the Suburbanized City

More than any other geographical context, the suburbanized city reflects the conflicts between women's and men's spatiality and temporality. Nineteenth century suburbanization and industrialization made women solely responsible for the home while men were engaged in work outside. A continuous trend from 1850 to 1920 has been shown of middle class women becoming more tied to their homes. This trend was associated with a decline in the number of servants and maids, so that women became housewives. The change was the result of several processes: suburbanization, which excluded women from central-city activities; social mobility, which reduced the importance and number of servants; and technological innovation, which included products designed to ease domestic work. In addition, the maintenance of a detached single-family home required more work than did an urban apartment, in which central services were provided.[46] All these factors were coupled with an ideology that placed women at home. This ideology also applied to single young women who lived with their parents in suburbia; young unmarried men preferred to live independently in central cities.[47] The lack of servants and the process of suburbanization thus caused women's temporality to become even more cyclical and their spatiality to become more constrained, since women became more tied to their homes. On the other hand, the factors of social change in the second half of the nineteenth century and in the early twentieth century yielded a contradictory trend in men's lives. Increased social mobility, technological innovation, and suburbanization intensified men's time and increased their daily spatial horizons. For women, as was seen, these same trends caused a deepening of the pre-capitalist pattern of cyclical time and small spaces.

The second half of the twentieth century has produced three stages in suburban development with regard to these differences between the sexes. The first period, lasting until the 1960s, was one of increasing geographical divergence between the locations of work and residence, so that central cities came to be symbolized with men and suburbs with women, though cities in general used to have a feminine connotation.[48] The increasing spatial size of cities through suburbanization and the introduction of expressways meant a further geographical separation between the work places of husbands and the strictly residential single-family suburbs or bedroom communities. The difficulty in bridging masculine and feminine spatiality and temporality have led some feminist writers to argue that suburbanization per se was created by men to oppress women.[49] This assertion seems difficult to prove, and indeed the economic and cultural roots of suburbanization lie elsewhere (see Chapter 3).

A second stage in twentieth century suburban development, from the 1960s on, saw the emergence of two simultaneous trends: the increased suburbanization of all economic activities, not just of population; and the increased share of women in the labor force.[50] The two phenomena are not necessarily interrelated, however. Though one might argue that suburbanization of employment could have reduced the pressures on women's temporality and spatiality, the low density of suburban development, especially in North America, prevented a reduction in distance and time of travel to work. Not only is modern suburbanization more extensive in nature, but any radial growth of a city results in a much higher geometric spatial expansion of metropolitan areas. Thus, women's spatiality and temporality with regard to commuting has been restricted, resulting in a lower flexibility of employment choice. The increased suburbanization of economic activities has not changed the temporal and spatial orientation of public transportation systems; they have continued to be oriented toward rush hours and employment foci rather than shopping, services and child care which are dealt with, to a large degree, by mothers and/or wives.[51]

The third stage in modern suburban development is telecommuting, a future development that will permit workers to work at home via telecommunications. This possibility has even been interpreted as bringing work back to the home, where it had been originally located before the industrial revolution.[52] On the surface, it looks like an excellent solution in terms of converging masculine and feminine temporalities and spatialities. Though some initial evidence shows that telecommuting could be attractive, especially for mothers, there are those who predict a role conflict between production and housework.[53] This last point should be emphasized, since the confined spatiality in traditional societies was total, meaning that housework, production, social life, and service provision all took place within a limited geographical range. Moreover, all these activities were not considered different from one another; thus, they shared similar tools and norms. Modern telecommuting, however, is just a spatial, not a directly social change, so that women's social worklife would be ruined if they worked at home. In addition, the role conflict between production and housework, which are so different from each other, might turnout to be stressful when located within the same geograhpical unit -- namely, the home, but requiring different temporalities.

Summary

Industrial and post-industrial societies have undergone changes in their spatiality and temporality, especially with regard to production and consumption. These changes, which accompanied capitalism, have been relatively easily adopted by men, but have been more difficult for women to adapt to, since their domestic roles require different modes of spatiality and temporality.

Spatial and temporal analyses of society and individual lives have been implicitly based on masculine conceptions and uses of time and space and have ignored the different spatiality and temporality of women. There are many urban-geographical topics that could be illuminated by taking into account the differentiation proposed here. Two of these should be of special interest. The first relates to the question of leisure and recreation. Are the temporal and spatial aspects of these activities similar for men and women, or are they also differentiated along sex? In the social sphere, a major question concerns the possibility of some future easing of women's tensions in regard to their dual spatiality and temporality. One of the "megatrends" observed for contemporary society relates to a social transformation from the more masculine hierarchical form of organization to the more feminine pattern of networks.[54] This trend could offer a promising avenue for further study of changes in temporality and spatiality along gender.

NOTES TO CHAPTER 4

1. On geographers and women's geography see Zelinsky et al. 1982. Two recent monographs on the geography of women are Mazey and Lee, 1983, and Women and Geography Study Group of the IBG, 1984. On time-geography and women see Palm and Pred, 1974; 1978.
2. Firestone, 1970.
3. E.g. Hartmann, 1976; Markusen, 1981.
4. Oakley, 1974; Zaretsky, 1976; Mazey and Lee, 1983.
5. Carlstein, 1982, pp. 403-405.
6. Boserup, 1970.
7. ibid., p. 16.
8. Zaretzky, 1976; Boserup, 1970; Mazey and Lee, 1983; Lowe, 1982, pp. 70-71; Chabaud and Fougeyrollas, 1978.
9. Toffler, 1981.
10. On the increased female sphere of activity conencted with urbanization see Wilson, 1979. On the increased separation between production and family life see Mazey and Lee, 1983; Miller, 1983.
11. See also Sargent, 1983; McDowell, 1983, p. 61.
12. Kellerman, 1985c.
13. Though, as Orme (1969) noted, one may identify biological and psychological cycles irrespective of sex, as well.
14. Mazey and Lee, 1983.
15. Shahar, 1983, pp. 157-158.
16. Toffler, 1981.
17. Cottle, 1976.
18. ibid., p. 11.
19. ibid., p. 14.
20. Rich, 1973.
21. Cottle, op. cit., p. 79.
22. ibid., p. 182.
23. MacKenzie and Rose, 1983, pp. 160-161.
24. ibid.
25. ibid.
26. Ericksen, 1977; Mazey and Lee, 1983.
27. See Palm and pred, 1974; 1978.
28. McGee, 1979, pp. 3-4.
29. Everitt and Cadwallader, 1972; Orleans and Schmidt, 1972; Appelyard, 1970; Pocock, 1976; Spencer and Weetman, 1981; Gilmartin and Patton, 1981.
30. Matthews, 1984; 1987, and others quoted by him.
31. Maccoby and Jacklin, 1974.
32. Sherman, 1978, p. 53.
33. Gilmartin and Patton, 1984.
34. Harris, 1978, p. 406.

35. Gilmartin and Patton, op. cit.; Women and Geography Study Group of the IBG, 1984, pp. 28-29.
36. Erikson, 1964, pp. 590-591. See also 1968.
37. On Erikson's ignoring of the influences of socialization see Lloyd, 1975.
38. Hopkins, 1980.
39. Nyborg, 1983.
40. ibid., p. 122.
41. Sherman, 1978; Harris, 1978.
42. McGee, 1979; p. 107; Hart, 1979; Saegert and Hart, 1978.
43. Saegert and Hart, 1978.
44. Munroe et al. 1971, p. 21.
45. For such a list, excluding time and space, see Wekerle et al. 1980, p. 27.
46. Jackson, 1985, pp. 48-49; Rothblatt et al. 1979, pp. 13, 37; Miller, 1983; Douglas, 1925, p. 85.
47. Douglas, op. cit., pp. 171-173.
48. On cities symbolized by men and suburbs symbolized by women see Saegert, 1981. On the female connotation of cities in general see Stimpson et al. 1981.
49. E.g., Wekerle, 1981.
50. On the suburbanization of economic activities see Muller, 1981; Kellerman and Krakover, 1986.
51. On women's travel to work in suburbia and its implciations see e.g., Fava, 1980; Eriksen, 1977; Mazey and Lee, 1983; pp. 26-27; Wekerle, 1981.
52. Toffler, 1981; Kellerman, 1984.
53. On mothers' work at home see Pratt, 1984. On the role conflict it may create see Salomon and Salomon, 1984.
54. Naisbitt, 1984.

CHAPTER 5

TEMPORALITY AND SPATIALITY IN ISRAELI SOCIETY: CHARACTERISTICS IN TRANSITION

Our attempts to uncover societal temporality and spatiality have taken us as far as the inner spatiality and temporality of men and women, in order to understand their collective time and space conceptions and behavior along gender. We move now to the opposite extreme of the societal spectrum, namely the nation-state, for which Israel will be used as a case-study.

Israeli society is, on the one hand, a new society, the first Jewish settlers of the modern era having come to Palestine only about a hundred years ago.[1] On the other hand, Judaism is one of the most ancient religions and cultures, having survived despite some two thousand years of exile from its national territory. It is, thus, of some interest to study two of the most basic societal aspects of Israeli society -- spatiality and temporality -- in light of its dual nature of old and new. This admixture becomes even more interesting when a recent increased influence of external, especially Western, notions of time and space in modern Israel is assumed to have taken hold. Several other unique circumstances typify temporality and spatiality in Israeli society. First, there is the constant reference to the Land of Israel within the Jewish tradition. Second, there is the secularism, sometimes extreme, of pioneering, modern Israel. Israeli society has also been characterized by rapid change, from totally traditional to socialist-oriented, and, eventually, to pluralistic with capitalist notions of time and space -- all within less than a century.

The following sections will attempt to discuss Judaic temporality and spatiality and the transitions in their characteristics in modern Israeli society. The major arguments will be that Judaic temporality and spatiality have been weakened in Israeli society. The first result of this state was that heavy political significance was given to these aspects in pioneering Israeli society. This period was later followed by a gradual change to economic-capitalistic notions of space and time; these dimensions were still coupled, however, with some specific political connotations.

Several discussions of societal time have emphasized its cultural and religious meanings, while others have noted its changing value under different political-economic systems.[2] Notable is the fact that a relative paucity of discussion exists on the political significance of time for the nation-state or for similar societal organizations.[3] As will be shown later, this very last facet is of considerable importance in the case of Israel, a society in which institutional ideology goes hand in hand with temporality and spatiality.

One exception to the rarity of discussion of the political dimension of time is Gross' work on temporality and the modern state.[4] In Chapter 2, mention was made of Gross' interpretation of temporality as the historical longue duree. Until the early nineteenth century, this temporality was under religious-traditional dominance; then the nation-state emerged and, among other things, transformed temporality into a national-political dimension. Currently, according to Gross, temporality is challenged by capitalism and its ahistorical nature. In the following sections, these three phases will similarly be identified for Israeli society. Beyond the obvious compositional differences between Israeli-Jewish and European-Christian societies, contextual and conceptual differences exist, as well. The transitions from one phase to another in Israel have been much faster than in Europe, since their beginning goes back to the turn of the nineteenth century only. The discussion of Israeli society will focus on both temporality and spatiality, and will refer to several aspects of the conception and use of time and space. As for temporality, both human and historical times will be treated. Our earlier discussions on urban society and gender differences dealt mainly with human time, the focus here on a national society requires an emphasis on historical time, as well.

The role of social values in the structuration process makes it evident that the ideology of both individuals and institutions constitutes an integral element of structurationist analysis, and the study of temporality and spatiality is no exception to this rule. Each social level in every production mode, it has been observed, maintains its own spatiality and temporality.[5] Giddens noted that each ideological group in society attempts to have its idea become generally accepted, which is an especially important point in the Israeli case.[6] In this respect, another writer went one step farther by suggesting a two-way relationship between the built environment and ideology, and thus between structure and practice.[7] Again, this is a point of special significance when Israeli society is analyzed. As will be shown later, it could explain, for example, the relationship between concepts of time and space, on the one hand, and urban and national expansion, on the other.

We now turn to analyses of Judaic, Zionist-socialist, and Israeli capitalist temporality and spatiality. The discussions will be separate for each approach and, within each, again separate for temporality and spatiality. Temporality will be elaborated by highlighting several aspects: the significance of time for each approach, cyclical and linear times, passive versus active time, extensive and intensive times, and time as a resource. Similarly for spatiality, the following aspects will be introduced: the significance of space for each approach, small/large spaces, passive versus active space, and extensive and intensive spaces.

Judaic Temporality

It is not the purpose of the discussion that follows to provide a comprehensive analysis of time in Judaism, since this may be found elsewhere.[8] Here, rather, an attempt will be made to identify some of the characteristics of Jewish time along the aspects that together make up temporality. By Judaism, we mean the Jewish religion in its traditional orthodox teaching, which served as the basis of both the religious and secular life of the Jewish people in exile until the modern era. As such, temporality as well as other aspects of life were sometimes geared toward life outside the national territory of the Land of Israel. Still, some of the bases of temporality originated during the Biblical and post-Biblical periods, before and during eras of Jewish permanence in the Land of Israel.

1.   The sanctification of time.   It has been pointed out that the "radical dimension affixed to time by Hebraic man" is a particular kind of historification of time: "the sanctification of time."[9] As in other religions, time in Judaism is divided into sacred and profane.[10] The sanctification of time in Judaism may be cross-classified by scale, tense, and degree. By scale is meant small time as opposed to big time; by tense, past, present, and future; and by degree, the different levels of significance given to different times. Small times are, for example, the Sabbath and the holidays. Their sanctity is expressed mainly in their recurrence in the present, but they also have important significance with respect to both past historical memories and future Messianic times, or big times. Big time consists, thus, of the long past and the longer future and has a different religious context than does the present. The past represents the national and universal memory, in which certain events have a unique holiness. The most important among these events are the Creation, Exodus, Torah Giving on Mt Sinai, the two Temples in Jerusalem, and their destruction. These holy events are not necessarily related to holy places. Creation involved the whole universe. Egypt, from which the Exodus took place, has no sacredness, and Mount Sinai became holy only for and during Torah Giving. It is only Jerusalem that enjoys a permanent holiness, and was so imbued with this status only after being chosen as the Temple site (in King David's time). (Certain traditions, though, attribute some holiness to the Temple Mount in Jerusalem as the site of the Sacrifice of Isaac and even as the site where Creation started.) The future is perceived through the yearning for the Messianic era and represents, thus, both national and religious hopes. In Judaism, then, "times becomes a conceptual 'field' for recording and deriving meaning out of the events of time. Time is the universal thread linking the events of time, and time thus becomes the nexus not of abstract t's, but of 'historified' meaningful events. The meaning of the events of time -- of a sanctified, historified time -- emerges as life's all consuming human endeavor for the Hebrew."[11] The sanctification of several time chunks in the present is of a different nature, since it requires the observance of specific practices. The several Jewish time scales and tenses and their significance are interwoven, a point that may best be demonstrated by the Sabbath.

The weekly holiday, the Sabbath, is the most important temporal unit of small time, despite or because of its weekly recurrence. It attains, together with Yom Kippur (Day of Atonement), the highest level of holiness. All religious norms, prescriptions, and customs relating to the Sabbath are geared toward the clearest possible demarcation of this day as a day different from the other days of the week. Thus, not only is work of any form prohibited, but the food, the table setting, house order, clothes, not to mention prayers, are all uniquely festive in order to ensure a relaxation of body and soul. The purpose is to achieve a unique partnership between the individual and his family and community, on the one hand, and between them and God, on the other. Observance has been termed "a leisure-work ethic."[12] Interesting for our discussion of temporality and spatiality is Heschel's statement that "the meaning of the Sabbath is to celebrate time rather than space. Six days a week we live under the tyranny of space; on the Sabbath we try to become attuned to holiness in time."[13]

The Sabbath is, however, related to big time, as well. As for the present, the reading of a weekly portion of the Torah on an annual cycle imparts a unique spiritual and historical meaning to each Sabbath. In addition, in Rosenzweig's phrase, "the Sabbath lends reality to the year."[14] The connection of the Sabbath to the past goes back to three major events, which are repeatedly mentioned during the Sabbath prayers and ceremonies: Creation, Exodus, and Torah-Giving. The Creation relates the origin of the Sabbath to God's Creation of the world in six days, and His resting on the seventh day. Exodus is mentioned as the start of Israel as a people; and Torah-Giving, as the source of the religious obligation regarding the Sabbath.

The Messianic future is also built into the idea of the Sabbath in several ways. First, since the observance of the Sabbath is considered one of the most important religious norms, it is believed that when all Jews observe at least two Sabbaths, the Messiah will come.[15] When this era comes about, the reconstruction of the Temple in Jerusalem, which will then be enabled, will change the form of the observance of the Sabbath through the resumption of Temple ceremonies. These ceremonies are therefore mentioned in the weekly Sabbath prayers.

2. Cyclical and linear times. Judaism adheres to both cyclical and linear time modes. Some students of time emphasize the linear modality of Jewish temporality, others the cyclical.[16] Cyclical time in Judaism relates to what we have termed small time. The calendar, with its weekly Sabbath, monthly New Moon, and annual festivals, represents cyclical time; the long past memory and indefinite future Messianic hopes represent linear time. The claim has been advanced that Judaism invented linear time by referring to time as historical time, within which God acts.[17] As such, one could argue that the first Biblical event, the Creation, marks the beginning of linear time. One may identify, however, several events that served as "pacemakers" for other events and for religious practices: These are, again, the Exodus, the Torah-Giving, and the two Temples.[18] Their role as pacemakers for the Sabbath has been demosntrated. On a more universal level, the Divine promise to Noah in the Book of Genesis to avoid future deluges may be regarded as yet another source of linear time, marking a unique, unrepeated event. Future linearity, too was introduced in the Bible through the Prophets' promises of national redemption and the reconstruction of Zion, which form the bases of Jewish Messianic hopes.

The differentiation between linear and cyclical times is usually helpful for two different areas of time study; namely, the cultural study of time and socio-economic analyses of production modes. When it comes to Judaism, this differentiation is slightly artificial. As we have shown, the most obvious symbol of cyclical small time in Judaism, the Sabbath, is strongly tied to linear and big past and future times. By the same token, the three major holidays, Passover, Feast of the Weeks (Pentacost), and Tabernacles, all have seasonal connotations of agricultural life and weather changes as much as they commemorate the historical events of the Exodus and the Torah-Giving. It might seem, therefore, as if these holidays have a very strong spatial meaning by virtue of the seasonal changes and the nature of agricultural life in Israel. These holidays, however, have been celebrated continuously through all the years of Exile and all over the world, their agricultural significance preserved in symbol and prayer only.

3. Passive time versus active time. "As a cause of events, the notion of sanctity suggests the image of time as an active, guiding force as opposed to the image of a mysterious line of time recording events."[19] A clear-cut differentiation between active and passive time modes in Judaism is difficult to make. Absolute time, represented by the calendar, is not necessarily passive or even completely absolute in Judaism. The first religious obligation directed by the Bible was to ensure the observance of Passover in the spring season.[20] This requirement eventually served as the basis of the man-made Jewish calendar, which employs the lunar year as its base, with adjustments for the solar year, as well.

Active time has received Jewish religious connotation through the ideal of national Sabbath observance, which, as we mentioned, would expedite the Messianic era. Time should generally be used to increase the portion allocated to Torah study and the fulfillment of other religious obligations. Secular time as an active resource was condemned by Rabbi Yehuda Halevi, the classical Hebrew 12th century poet, in what is considered the shortest classical Hebrew poem: "Slaves of slaves are the slaves of time - only the slave of God is free." The idea of active time-use for the construction of the Land of Israel as a religious obligation had to be left for religious Zionism, which came into being only in the second decade of the present century. Ultra-orthodox Jews, who oppose religious Zionism and Zionism in general, see the two active time modes of Torah study and the construction of Israel as contradictory. Thus, according to their interpretation, the reconstruction of the Land of Israel is not an active religious use but a profane use of active time. For this group, active time has a more restricted connotation of Torah study, prayers, and dealing with other religious obligations. These two approaches to active time-use are reflected in a Talmudic debate between Rabbi Yishmael and Rabbi Shimon Bar Yochai. The first advocated a combination of agricultural work and Torah study, while the latter called only for the study of Torah.[21]

4. Extensive and intensive times. Active time maybe used in two different modes -- namely, extensively and intensively -- and both modes take place in Judaism. The Sabbath is the peak of extensive time, while Friday is the peak of weekday intensive time, when preprarations for the Sabbath form the center of activity. By the same token, the seventh year in the cycle of agricultural years, the fallow year (or Sabbatical year), is an example of extensive time in production, as a special burden of intensity is placed on production during the sixth year. The idea of extensive and intensive times goes back to the Creation, which took six days and was followed by the Sabbath. Thus, both extensive and intensive times are historically ancient and Divine.

5. Time as a resource. Time is considered in Judiasm, as a basic, extremely important dimension and a framework for religious practice, memory, and hope. Time may also be interpreted as a religious resource in the intensive uses made of it on certain days and years. It might be viewed this way with regard to the study of Torah, for which as much time as possible should be allocated. It is difficult, however, to see a specific, clear Judaic interpretation of time as a political or economic resource. On the other hand, time is considered to be an important resource in general. The stealing of somebody's time has been compared in its severity to stealing a material object.[22] Perhaps it is even more severe, as this stolen time cannot be returned. This last notion of time as a "lost thing" served the Jewish Sages when they permitted certain types of work during the holiday weeks of Passover and Tabernacles.

Judaic Spatiality

Judaic temporality has been shown to be of basic importance in daily Jewish practice and belief. It has brought one writer to declare that Judaism is a "religion of time aiming at the sanctification of time [whose| main themes of faith lie in the realm of time."[23] The role of spatiality in Judaism is more restricted compared to that of temporality. This may largely be due to the two thousand years of exile, which have brought Judaism to think and act within the "unnatural" spaces of the Diaspora. There are those, however, who identify the roots of the more favored temporality in the Bible itself. "The Bible," Heschel wrote, "is more concerned with time than with space. It sees the world in the dimension of time. It pays more attention to generations, to events, than to countries, to things; it is more concerned with history than with geography."[24] As another writer saw it, in this same view, the Judaic "holy places were a celebration of events not a reverence for primordial holiness per se."[25]

The discussion below will briefly address the major aspects of Judaic spatiality: the sanctification of space, small space/large space, passive space versus active space, extensive and intensive spaces, and space as a resource.

1. The Sanctification of Space. The Divine promise that the Land of Israel would forever be the national territory of the Jewish people was -- and is -- a very basic element in Judaism. It became an integral part of the Jewish experience.[26] The Rabbis regarded the universal stories in the Book of Genesis as a Divine justification for the giving of the Land of Israel to the Chosen People, the seed of Abraham.[27] Thus, the Land of Israel is God's land; it has a basic, eternal sacredness, independent of time and the deeds of the Jewish people. This basic sacredness, however, has been elevated by additional levels of holiness that were added at different historical times, with their concomitant changes in geographical boundaries and religious significance. There is the "first holiness," originated by Joshua and the Children of Israel in penetrating the Land. The "second holiness" dates back to the Jews' returning from exile in Babylon and their constructing the second Temple. A "third holiness" will be established in Messianic times.[28] The establishment of this last holiness is

dependent to some degree on the conduct of the Jewish people. Contrary to what may be thought, the term "holy land" seldom appears in the Bible, so that one is left with the impression that it is derivative.[29]

The survival of Judaism outside the Land of Israel was coupled with the increased importance attributed to time in contrast to space. Spiritual attachment to the Land of Israel has remained very strong, however, as reflected in prayers, festivals and Rabbinical teachings. Daily spatial practices were, necessarily, limited in the Diaspora. Some students of Jewish space have argued that the Torah served as a substitute for "moveable territory" or that the synagogue could be regarded as such. Others considered the strong temporality to be a substitute for the people's weakened spatiality.[30] Even in ancient times, however, major events in Judaism occurred outside the Land of Israel. Abraham came to Israel from the East, and the Children of Israel came from Egypt. The Torah-Giving occurred in the Sinai desert.[31]

The sacredness of space finds another, and completely different, expression in Judaism, through one of the names of God -- "The Place" (Hamakom). This appellation does not appear explicitly in the Bible although one could argue that it was attributed to a relationship with God by Abraham, Jacob, and Moses.[32] This name was first used around 300 B.C. by Simon the Just. Some scholars believed it presented Persian or Hellenistic-Alexandrian influences.[33] In the third century A.D., after earlier starts, a contamination process gradually replaced this name with another, "The Sanctified One, Blessed Be He" (Hakadosh Baruch Huh). Since then, the name "The Place" has been used only for certain specific blessings. The Rabbis did not know the origin of this name even in the third century A.D.[34] The Talmudic Rabbi Ami pointed to the problem of its origin and averred that God was called as such "because He is the place of the world"; He quoted another Rabbi (Yossi Ben Halafta) who said, "He is the place of His world, and His world is not His place."[35]

We have gone into some detail regarding this issue, since it is strange that the extremely abstract notion of God in Judaism would permit the use of a definitely spatial, and thus limiting, name. Whether this really is the case and whether Rabbi Ami's answer was meant to defend the use of the appellation has been debated.[36] In any case the nature of his answer is a posteriori and apologetic. One modern interpretation of "The Place," that it is "mainly indicative of God's ubiquity in the world and can best be translated by 'omnipresent,'" is interesting since it uses a temporal term to translate and interpret a spatial one.[37] A different interpretation states that "if the universe is unthinkable without a space frame (and this is indeed, the crux of Kant's a priori concept), so much more so is the Jewish world incomprehensible without an all-embracing God.[38] Kabbalistically, it has been demonstrated that squaring the numerical values of the Hebrew letters of God's name yields the same number as does the summation of the numerical value of the word "place" (makom).[39] In any case, if this appellation is indicative of anything, it shows that space has a very basic sacredness in Judaism when relating to God, not necessarily to a specific territory (i.e., the Land of Israel).

One more interpretation goes as follows.[40] Jerusalem was called "the place" as is apparent from a reading in Deuteronomy of the obligation of the first fruit and from the high court (Sanhedrin) being termed "place".[41] Thus, the name of the holy city is used as a metonym for God's name, an interpretation that blends the sacredness of God with that of a holy territory, namely Jerusalem.

2. Small Space/Large Space. Within the Land of Israel, the smaller the spatial unit, the higher is its level of sacredness. Thus, the Land of Israel is holier than other countries, the city of Jerusalem holier than the rest of the country, and the culmination is a small room within the Temple known as the Holy of Holies.[42] The destruction of the Temple and the city of Jerusalem in 70 A.D. did not prevent the Tannaim, the authors of a part of the Talmud, the Mishna, in the second century, from describing the edifice and the city in full details of function and holiness in what one writer has called "a map without territory."[43]

Within and outside the Land of Israel, the synagogue is imbued with a special holiness. In it is the Holy Ark, which has a higher sacredness, and the Torah Scrolls stored in the Holy Ark have an even higher level of holiness.[44] On the other hand, the notion of "holy places" in Judaism is doubtful, with the exception of the Temple Mount. Thus, the grave of Moses and the location of Mount Sinai have not been known in order to prevent site-specific worship in these places. Nevertheless, the holiness of the Land of Israel has a changing historical context, and the same is true with regard to the sacredness of the Temple Mount and the city of Jerusalem.[45]

3. Passive Space Versus Active Space. The Land of Israel may be regarded as passive space, since it is the "natural" arena for Judaism: The territory is the homeland of the Jewish people. Even this "given" aspect, however, cannot be considered static when the changing holiness of the Land is taken into account.

Active or relative space in Judaism is reflected in the ways space should be organized both within the Land of Israel and in Jewish communities in the Diaspora. Thus, the obligation to live in the Land of Israel is considered the most important religious requirement although the Rabbis could not agree on the applicability of this obligation in current, pre-Messianic times.[46] In addition to the religious significance of residing in the Land of Israel, the development of the Land is a religious requirement.[47] On the smaller, community scale, special laws apply to the location of the synagogue and the cemetery.[48]

The active role of land, from a Judaic viewpoint, is also expressed in laws pertaining to agricultural production within the Land of Israel. These "land-dependent obligations," as they are called, have the aim of regulating agricultural production itself and the distribution of yields.[49] For example, they control the temporal dimension of production through the fallow year as well as the spatial order of crops on a given unit of land. They also regulate social order so that the poor and the priests (the latter, a landless caste) will receive their share of the yields obtained by land owners.

Space as an independent dimension has a universal, worldwide, importance in Judaism only at the micro scale of the home, the synagogue, and to some extent the community. Space at the macro scale of national territory and agricultural production has significance only within the boundaries of the Holy

Land. Time, by contrast, is a universal dimension without differentiation between the micro and the macro.[50]

4. Extensive and Intensive Spaces. The notion of a more intensive use of space for specific religious purposes does not exist in Judaism. Regulations with regard to the intensity of land-uses are directed toward the achievement of other objectives. For example, the distance between houses should eliminate any possible obstruction of privacy.[51] Judaism's near lack of an attitude toward space-use and intensity is, thus, rather different from its special organization of time into intensive and extensive periods as a religious requirement per se.

5. Space as a Resource. Space is regarded in the Bible as a basic economic resource, especially concerning agricultural production and the division of land among families and tribes. This view is expressed in such laws as the Jubilee, which requires the return of land to its original owners, and the laws of the fallow year.[52] In addition, space already had an exchange value in ancient times for non-agricultural uses. A very manifestation was the purchase of a gravesite by Abraham for his wife, Sarah. The Land of Israel, furthermore, is described as a plentiful agricultural resource and as a Divine gift.[53]

The Talmud discusses the details of laws regarding the sale of land, but these do not necessarily have a "religious" loading and their major purpose is to arrange for proper land exchanges. The significance of space as an economic resource, however, was much higher than that of time. The mass commodification of time came only with the industrial revoluton, while that of space was ancient.

Judaic Temporality and Spatiality: A Summary

Both time and space have been shown to have basic significance in Judaism in various modes and ways. There is an eternal holiness attached to both time and space. Nevertheless, the degree and significance of the holiness of space, especially the Land of Israel and Jerusalem, have changed over time; on the other hand, the sacredness of times, such as the Sabbath and the holidays, has been considered eternal and unchanging. This difference is not only theoretical or philosophical; it is of important practical value, since it determines changing religious obligations and practices regarding space. It also assures a continuity in temporal practices, regardless of location. The difference has led several scholars to the view that Judaism is a tradition of time rather than of space.[54] Others present a different argument to reach the same end; although the Land is inseparable from the Torah, they say, "Judaism is not a territorial religion: the Land is not of the essence."[55]

The important distinction between eternal times and changing spaces might stem from the long periods during which the Jewish people were separated from their homeland. It would, thus, be of much interest to investigate the attitude toward time and space of the new Israeli society, which marks the reunion of the Jewish people and their national territory. The increased value given to time in

Judaism has had a deep impact on Jews, even when not practicing Judaism and when dealing with non-jewish subjects. As mentioned in Chapter 2, three famous, modern, Jewish-born thinkers, Bergson, Freud and Proust, all attributed more importance to time than to space and opposed the spatialization of time. This shared view has been explained as being a consequence of their Jewish heritage.[56]

In terms of the "secular" economic value of time and space as resources, space is more important than time in Judaism; indeed, space was already viewed as a commodity in ancient times. This attitude might be attributed to the socio-economic conditions within which Judaism developed; namely, in an agricultural society, in which space was more valuable as a resource than time. Time gained in importance as a resource in modern Israeli society. This increased value of time has been secular-economic in nature, and not necessarily based on a renewed Judaic perspective on time as an economic resource.

We come now to an assessment of time and space in pioneering, modern Israel. It is important to point to a certain difference in emphasis between the discussion of time and space in Judaism, on the one hand, and in the new Israeli society, on the other. An evaluation of the attitudes to time and space in Judaism meant going into an analysis of a culture. As such, the conception of time and space has sometimes been emphasized more than their uses. In the upcoming sections, in which society rather than culture will be the focus, the modes of time and space uses will be increasing subjects of reference. As will be apparent from the discussion to follow, this difference between conception and use is not always as distinct as it is made here. It points to a two-way relationship between the societal conception and uses of time and space. The difference will once again attest to our view that the definition of temporality and spatiality should consist of both the conceptions and the uses of time and space, respectively.

Temporality and Spatiality in Israeli Society: General Trends

The issues to be addressed here relate to temporality and spatiality in Israeli society as a Jewish society that experienced renewed contact with its national spatial framework after many years of separation. These issues received an additional importance, since the modern settlement of the Land of Israel has been accompanied by a process of secularization of thought and deed. This happened even though most of the pioneering founders of Zionism originated in the traditional Jewish society of Eastern Europe, with its strong temporal and weak spatial contexts.

Our major objective in the coming sections will be to show the increased importance attached to spatiality and the strong political connotations given to both spatiality and temporality in pioneering Israeli society. We shall argue further that in current Israeli society, both aspects have taken on a more economic-capitalistic meaning within an urban context. These arguments, however, should not lead to any assumption that the traditional Jewish interpretations of temporality and spatiality have been repalced, or that they have disappeared, at either the individual or the societal level. Rather what they propose is that a change of accent has occurred in spatiality and temporality in Israel. The several elements comprising spatiality and temporality will again be reviewed, once for the pioneering Israeli society and a second time in more condensed form, for contemporary Israel. This time, the order will be reversed and spatiality will be discussed first followed by temporality.

Spatiality in Pioneering Israeli Society

"Pioneering Israeli society," in our definition, refers to the era between 1904 and 1957. The year 1904 marks the beginning of the "second Aliyah," the immigration wave that initiated Israeli socialism. It was followed by the establishment of the first collective group, which in 1909, settled the first Kibbutz. Though "modern" agricultural settlements already existed since the 1870s (the Moshavot of the "first Aliyah"), the Kibbutzim were of a different nature. The Kibbutz was a collective settlement, and its establishment marked the beginning of an era of both massive settlememt of the Land of Israel and an ideological attempt at imparting a socialist character to the new Jewish society in Israel. Despite the strong political impact of socialism on Israel's emerging societal life, other political camps developed, as well. Among the latter were the Revisionists, with their nationalistic orientation; the urban, capitalistic middle class; and the religious Zionists, who tried to bridge between socialist Zionism and traditional Judaism. In addition to these new factions, there already existed also an ultra-orthodox population, which adhered to religious Judaism in its strictest sense and objected to any new "external" influences. Socialist Zionism being the core of the Israeli pioneering society, will receive the emphasis in our discussion of spatiality and temporality; but comparisons will be made with the other parts of the emerging Israeli society.

The end of this pioneering era came in 1957, following the Sinai War of 1956; the former year is considered a landmark in the modernization of Israeli society, though signs of the upcoming change could already be seen in the early fifties.[57] As will be pointed out later, this does not mean that a sudden change in spatiality and temporality could be observed in 1957. The transition was gradual, with additional strong inputs provided by the Six Days War in 1967.

1. The Significance of Space. Zionism may be interpreted as a spatial ideology par excellence. Although every national ideology has a strong territorial component, this aspect in the case of Zionism has received a special status because of the unique circumstances of Zionist activity in the Land of Israel. The return of Jews to the Land of Israel after a separation of two millenia had an enormous emotional impact on the attitude toward the Land itself, though most Zionists and Zionist parties did not interpret it in religious ways. The poor physical condition of the country and the struggles with the Arabs, the Ottoman Turks, and later the British Mandate authorities required the evoluton of a very strong attachment to the Land in order to develop a growing, firmly based

Jewish society. Thus, some writers prefer to emphasize the role of Zionism as a national-political movement separated from Jewish theology.[58]

The Zionist attitude toward the Land of Israel could be interpreted as a political and cultural "holiness" that was manifested in many ways by different groups. This view was not shared by the ultra-orthodox Jews who lived in Palestine before its Zionist development, and who continued to adhere to their strictly religious spatiality, in which there was no place for political aspects. The religious Zionists, on the other hand, attempted to integrate religious with political spatiality. Their basing the political spatiality on the religious, an exercise totally rejected by the ultra-orthodox circles, did provide, however, for a genetic link between the religious and political spatialities. This is not to say that secular Zionism, of whichever form, did not base its doctrines on Jewish sources. Rather, its interpretations of these sources and the resulting attitude and practice were not religious. For example, many holidays received a renewed accent on spatial-agricultural elements. Thus, the Feast of Weeks was celebrated mainly as the festival of the First Fruits, the fact of its also being the celebration of the Torah-Giving being ignored.

2. Small Space/Large Space. The attitude toward small and large spatial units differed among the several Zionist parties. At one end was middle-class Zionism, which referred to land as an economic resource to be used mostly for urban development. It was middle-class pre-Zionist East European Jewry that had initiated the first modern rural settlements (Moshavot) during the First Aliyah (immigration wave) during the last quarter of the nineteenth century. There was a sharp decline in the continued contribution of this sector to rural development with the start of the socialist Second Aliyah. Thus, between the years 1878 and 1904, 17 Moshavot were built, whereas between 1904 and 1948, only 33 Moshavot were added, compared to the 149 Kibbutzim and 87 cooperative Moshavim that were constructed during the same period. Since rural development was a major Zionist tool for expanding and holding national large space, the attitude to big national space, although favored by middle-class Zionism, was weaker. On the other hand, middle-class Zionism had a crucial role in the development of the urban sector through private purchases of land and the construction of housing and industries. The rapid growth of Tel Aviv, Jerusalem, and Haifa was greatly due to the efforts of middle-class Zionism. The labor movement generally opposed this trend of urbanization.[59]

Middle-class Zionism was closer in its outlook to the rightist Revisionists' approach of achieving Jewish sovereignty through political struggle, though the former did not share the latter's concept of armed struggle against the British administration and the Arabs. The emphasis of middle class Zionism on urban development and on land as an economic resource may be interpreted as a focus on small space. At the other extreme regarding spatiality was Revisionist Zionism. Although it also found a common language with middle-class Zionism on economic issues, Revisionist Zionism adopted a large-scale spatiality: a belief in the necessity of ensuring a Jewish state on both banks of the Jordan river. The achievement of Jewish sovereignty in the Land of Israel, whether by political or military means, was the Revisionists' most important, most immediate objective. Vladimir Jabotinsky, the founder of Revisionist Zionism, had declared:

> Settlement in its fullest sense, settlement of masses of people is impossible without a political domination over that territory that you want to settle. There is a long list of conditions for settling a territory which can be treated only by political domination. There must be the right laws, the right conditions for transportation, an undisturbing but assisting adminsitration. All this can be achieved only through a charter, only when the authority and the political power are in the hands of the settler himself.[60]

The Labor Zionists, both secular and religious, tended to adopt both small-scale and large-scale spatialities. On the one hand, they, too, emphasized the need for Jewish sovereignty in the Land of Israel, which represents large space. This aim was coupled, however, with a view of rural-cooperative settlements as the chief means for its achievement. As such, the ideal of small settlements scattered over desired areas throughout the country represents small-scale spatiality. This view manifested itself in tactics of piece-by-piece land purchase and cultivation; favored as slow but sure deeds on the way to independence ("a dunam here, a dunam there," as the popular Hebrew phrase had it). The mix of large-scale and small-scale spatialities may best be demonstrated by the words of Yitzhak Tabenkin, one of Labor Zionists' strongest proponents of settlement activities:

> Without settlement there is no nation. Our nation is a fruit of settlement. Settlement means identity with the country's landscape. Our nation is an Israeli nation. The country is the mother of the nation. Only the nature of our settlement in the country determined our character as a nation. This determined our language as well.[61]

Tabenkin had this answer for the Revisionists' ideology:

> There was a party in Zionism that wanted that the settling of the Land of Israel will be done in a way of a settling government, by the Jewish State. Surely a state can settle, but first of all, it was necessary to create the basis for the state, the elements for sovereignty and economic and settlement independence. Political plans only do not make for settlement.[62]

3. Passive Space Versus Active Space. Both Labor and Revisionist Zionism related to space actively, while middle class Zionism adopted a more passive view of space.[63] As we have seen, these groups exhibited major differences with one another with regard to spatial scales and a passive versus an active space approach.

The Labor Zionist approach, the leading Zionist thought and action in pioneering Israel, made the active view of space into a basic social value, with a strong geographical connotation. Thus, Israelis believed in geographical action as a crucial determinant of political life, and expressed this ideology in settlement construction and agricultural cultivation. This point is well demonstrated by the still current Hebrew expression, "to create (or fix) facts on the land." Although the expression was, and is, widely used in both Zionist writings and daily language, it cannot be found in any dictionary. Not only did the syntax of this idiom have no earlier equivalents, but part of the expression, "to fix facts" is defined in one dictionary as "ordering things as they are," which carries an a posteriori connotation for things already in existence.[64] By contrast, "to create (or fix) facts on the land" has an extremely a priori connotation, of making facts per se. This "making" of facts is in fact expressed in strong wording: "creating or fixing" symbolizes not only the new but also the strong; "facts" refers to complete deeds that have to be taken as given; and "on the land" points both to the public character of the deeds and to their geographical context, settlement acts. This modern Hebrew expression is much stronger than both the concept of "presence" on the land, which is typical of new societies, and even the concept of faits accomplis, attributed to Jewish settlement activities.[65] This is so because "to create facts on the land" refers to an action and to the people who create the facts rather than to the more passive and a posteriori existing facts themselves. It containts, thus, an element of activism, which is another characteristic of pioneering Zionism.

Language, some writers have argued, can play an important role in the development of places.[66] By the same token, we may argue here for the role of language in creating and fostering spatiality and temporality. The active attitude toward space, as expressed in the Hebrew idiom of establishing facts on the land, might be related to five elements. First is the notion of possession (Hazaka), which in both the Jewish and Ottoman codes assigned land ownership to someone who either held a piece of land or cultivated it for some time. A second social element reflected in this idiom is persistence and insistence, which have been considered Jewish characteristics for thousands of years.[67] Third is the Jewish need to settle lands fast, sometimes overnight, under both the Ottoman and British regimes, for the Ottomans did not destroy houses that were already roofed and the British were reluctant to remove existing settlements. A fourth factor behind this expression is the issue of national security, which was interpreted in Zionist ideology as the need for more civilian settlements.[68] Fifth is activism per se, a characteristic of Zionism in both pre- and post- statehood.[69]

4. Extensive and Intensive Spaces. The major contrast here is between labor Zionism and middle-class Zionism. The first advocated an extensive space use at the national level by spreading small settlements all over the country. As one Labor Zionist leader stated, "Our borders are not secured by their political recognition. Nor will they be secured by any army - such a big army we will not be able to establish - but by a big and fast movement for the seeding of working settlements on the borders of the country and its deserts."[70] Nevertheless, middle-class Zionism, on the other hand, used space in a rather intensive form.

This was the case at the urban level, the major cities of Israel having been built and developed as compact, dense entities.[71]

5. Space as a Resource. The Zionists of all groups considered space foremost as a political resource in both its small and its large scale. Land purchase and use were usually not done for strictly economic reasons, but for political motives. As long as land was held in non-Jewish hands, it was assessed only economically so that a price could be negotiated with the sellers. Once the land was transferred to Jewish hands, it achieved an ideological-political value. Purchases were now termed "land redemption." Public or institutional ownership of land by Zionist organizations thus served as a substitute for sovereignty.[72] Nevertheless, middle-class Zionism still referred to land as an economic capitalist resource in an urban context.

Temporality in Pioneering Israeli Society

Pioneering Israeli society, even though it functioned for almost half a century without sovereignty, may be considered a nation-state society. The Zionist movement was established for and geared all its activities toward the goal of the creation of a Jewish state in Palestine, and except for the ultra-orthodox, Zionism was an ideology shared by almost all Jewish immigrants to Palestine. Finally, the British Mandate itself extended relatively wide autonomy to the Jews, especially in the areas of settlement, education, and internal politics. A discussion of temporality in Israeli society might begin, therefore, by putting this society in the context of wider trends shown by nation-states.

"The state," said Gross, who defined temporality as longue duree in the nation-state, "is presently the chief institution of the modern era determining our 'representation' of the past."[73] This taking over by the state of temporality that had been shaped by religious tradition was also true of pioneering Israeli society. In Israel, as in the West, this meant using education to put more accent on national than on religious holidays (small time) and on historical events (big time). It further meant the weakening of religious institutions. In both pioneering and modern Israeli society the socialist and the state ideologies, respectively, attempted to impart new meaning to the originally religious Sabbath (Saturday) and holidays, which they considered to be national times. The clash, therefore, between religious institutions, on the one hand, and state and secular organizations, on the other, regarding temporality (and to a lesser degree regarding spatiality, as well) has been more central than in Western societies.

The Western nation-states, it has been said, "were not interested in temporality as such.... All they were concerned with was securely establishing the authority of the state over other kinds of institutions that stood in their way."[74] As we shall see Israeli pioneering society was different: it was interested in many facets of temporality per se. This stemmed from the strong deep significance of temporality in traditional Judaism. Paradoxically, conflict and change occurred under circumstances of a spatiality that was more important than temporality.

1. <u>The Significance of Time.</u> Time was actually subordinate to space in pioneering Israeli society, since time was conceived as a major dimension and resource for the achievement of space and spatial goals. On the other hand, the unique role of small time in traditional Judaism was diminishing in a society that was increasingly becoming based on secular ideas. This is not to say that Judaic time disappeared. Not only was there a religious minority within Israeli pioneering society, but the Sabbath and the traditional holidays were instituted as public holidays for the Jewish population. The significance given to the Sabbath changed, however, from a religious day in celebration of a unique time to a weekly rest holiday with or without several traditional customs.

The attitude toward big time changed as well, from relating to it as a religious dimension to conceiving of it as a secular element. Thus, for example, the Biblical notion of Jubilee was reinstated in a secular form. According to Biblical law, land should return every fifty years to its original owner/tribe. This obligation, it was reasoned, obtained by virtue of the Divine ownership of land.[75] Jubilee has not been observed since the destruction of the First Temple, because the calculation of Jubilee cycles was not preserved, following the expulsion of the Jewish people from the Land of Israel. The Jewish National Fund, which was the major vehicle for the purchase of land in pioneering Israel, did not sell land to private users; instead, it leased it, for 49 years. This policy was later adopted by the State of Israel. The difference between the two systems lies, of course, in the origin of land ownership: in Biblical law, it was Divine; in pioneering Israel, it was the nation or the society.

Zionism related to time in specific ways with regard to both past and future.[76] The preferred times from the historical past were the Biblical and Mishnaic eras during which strong daily bonds could be found between the Jews and their land. This preference found expression in the process of place name-giving in Israel. Thus, about 40 percent of all place-names in Israel are of Biblical-Mishnaic origin, whereas only 16.5 percent reflect Zionist motives. The two millenia of life in the Diaspora were given little recognition in place-names.[77]

The future, for its part, has been looked upon, especially by the Labor movement, as an active dimension for the development process of the country. This view was contrary to the more passive, spiritual view of big time by traditional Judaism. The difference between the passive-religious and active-Zionist conception of time has caused special tensions between religious Zionists and religious non-Zionists.[78]

Several temporal values have been mentioned so far; namely, the subordination of time to space, the political connotation given to time, its reduced importance when compared to Judaic time, and the differentiated attitude to the long Jewish past. Taken together, they seem to present a spatialization of time. This spatialization means that time was conceived of and used in a strongly linear and active manner, normally typical of spatiality. Temporality has been linear, since the Jewish past was considered an early building block for present and future national life in the Land of Israel. Thus, any interruption in this line, such as long years of exile, have less importance. Time was active in the sense that the attitude toward the historical past was not that of a series of given deeds and developments, but of events with differentiated lessons and meanings for the present and future. For example, the element of struggle that characterized the ancient Hebrews was carried into the Jewish struggle with the Arabs. Thus, the impacts of topography and nature on ancient battles were carefully studied in light of security needs. In addition, Biblical agricultural motives were revived. Time, then, was conceived of as active for a live national framework for the construction of the Land.

Another, more "external" aspect of time as a political resource in pioneering Israel related to its role in the Jewish-Arab conflict.[79] Until 1948, the Arabs wanted to preserve the status quo while the Jews were interested in change. This was interpreted as time favoring the Jews. After 1948 and until 1967, the situation was reversed in that the Jews wanted to keep the status quo until the Arabs recognized Israel, while the Arabs were interested in a change, specifically the destruction of the new State. Between 1967 and 1977 the attitude toward time reversed itself again, such that time was considered by the Arabs to be working in favor of Israel, which was interested in maintaining a status quo in the captured territories. The situation for Israel after 1967 was similar to that of pre-statehood from a spatial perspective, in that Israel controlled territories without the power of sovereignty. It was different, however, from a tempoeral viewpoint, in that the status quo had become important. After 1977, when the Likud party -- which combined Revisionists and middle-class Zionists -- came into power and peace negotiations with Egypt started, time was considered by the leaders of Israel again in favor of Israel, which wished to settle Judea, Samaria and Gaza as fast as possible. Whereas the earlier interest in keeping a status quo (1948-1967) meant using time passively, time was used actively, after 1977, as it was before 1948, for settlement activities.

2. Extensive and Intensive Times. Revisionist Zionism preferred an intensive time-use, while Labor Zionism advocated an extensive use of time.[80] The Revisionists wanted immediate Jewish sovereignty in Palestine, while the Labor movement wished to see a social revolution and an extensive settlement of the territory, both of which required an extensive time-use. The extensive time-use advocated by Labor was interrelated with its extensive space-use, reflected in scattered rural settlements across the country. After 1937, however, Labor's attitudes to both time and space changed because of the evolving struggle between Jews and Arabs (1936-1939), which made necessary a more intensive use of time, for a faster settlement-pace of Palestine. The political atmosphere also changed as a result of the persecution of the Jews in Europe. These events required spatial planning for the future settling of Palestine, and further called for faster settlement and political action toward the achievement of statehood, even if social goals had not yet been achieved. Thus, between 1907 and 1935, some 43 Kibbutzim were established; however, between 1936 and 1939, the number of Kibbutzim was almost doubled to 80; and by 1948, there were 149 of these collective settlements.[81]

3. Time as a Resource. Time started to be an economic as well as a political resource since the 1920s. This aspect of time was limited, however, to middle class, urban Zionism, which gave time its capitalistic importance as a major production resource, especially in manufacturing. This significance of time, however, was limited in a society having socialist predominance, limited industrial production, and financial services still in their early stages of development.

Temporality and Spatiality in Modern Israel

The period since 1957 has witnessed tremendous changes in Israeli society in several respects. The national economy became more mature. Agriculture moved from production mainly for domestic consumpton, with the exception of the exporting of citrus fruits, to the most modern, export-oriented production of all fruits and vegetables. Industrially, the economy changed from an emphasis on light industries, such as cheap textile and naive attempts to achieve self-sufficiency in all industrial products, to sophisticated, specialized, export-oriented industries, with a strong high-tech sector. The 1957-1967 decade, between the Sinai and Six Days wars was typified by the massive modernization of agriculture, while during this period the change in industrial production was still incubating. Massive industrial change came after 1967, with the prosperity and economic growth that followed the Six Day War. Since 1977, industrial expansion has become more specialized, and a boom has been witnessed in the tertiary services sector, public, private, and financial.[82]

The 1960s and the 1970s were also marked by a continued concentration of Jewish population in urban places. Tel Aviv became a metropolitan area of over 1.5 million people, while Jerusalem and Haifa each had a population of 400,000. This urbanization became an important factor. Israelis have persistently preferred urban environments for their residences, despite on-going governmental plans and incentives for population dispersal.[83] The cities, which have traditionally had a large private sector, have thus contributed to the reshaping of spatiality in modern Israel from socialist and a heavily political bent to a society that is more capitalistic oriented.

The economic changes have brought about a constant increase in personal income and the increasing exposure of Israelis to the West, in both production and consumption. Production of goods and services has followed the capitalist mode even in the public and labor-owned sectors. Economic considerations of productivity, efficiency, and profits replaced the Zionist-socialist value of production for the sake of production. The earlier socialist values had called for the development of industry in Israel at any costs, thus providing employment and an industrial infrastructure. Consumption patterns, with increasing income, have followed the example of the West, with the mass consumption of private cars, major appliances, and growing international travel. The new modes of consumption have been influenced by the mass media, especially T.V. The changing modes of production and consumption were coupled with increasing governmental efforts to construct modern road and communications networks.

Thus, modernization in Israel, in both production and consumption, led to the capitalist notions of intensive time-use and more extensive space-use.

These transitions occurred within a short time and meant placing economics at the center of the national agenda. The new emphasis, however, did not result in a disappearance of political Zionism as a central aspect of Israeli societal life. As we previously noted, the continued struggle with the Arabs and the conquest of territories in 1967 have, to some degree, turned the time-wheel to pre-statehood attitudes with regard to settlement problems and needs. There were now, once again territories not under Israeli sovereignty, they were formally under military occupation -- but that Israel wished to retain, at least partially. Some of the territories (Golan, Jordan Valley, Yamit region), were regarded as crucial for security reasons; others (Judea and Samaria), were perceived by parts of the Israeli society as indispensable for a combination of reasons: security, national- Zionistic, and religious. This unique blend of politics and economics in modern Israel caused a similarly unique merging of temporality and spatiality. As we shall see, the country's politics and economics since 1973, and even more so since 1977, were accompanied with a renewed religious temporality and spatiality in parts of Israeli society.

Temporality in Modern Israeli Society

The major recent change in temporality in Israeli society has been the increased economic importance given to time as a production resource. This change has been manifested in tight production and marketing schedules for both manufacturing and services and, thus, in a stricter measuring of working time for payroll and for production costs. Increasing incomes also made the financial time of both individuals and financial institutes an important element of daily life. In the decade from the mid-1970s to the mid-1980s, this element received an amplified role, even beyond that typical of mature capitalist societies, because of the high inflation rates that prevailed.

As mentioned above, the notion of time as a production resource had been adopted earlier by middle-class Zionism and considered peripheral at the time by the dominant Labor Zionism. It has now become central in Israel's economic life. The trend actually started under the Labor government, with the modernization of industry, and intensified with the change of administration in 1977. The evolution of this notion of time may, therefore, be regarded as a process of the nesting of economic capitalist time within political-socialist time. It was similar in process to and concurrent with the evolution of settlement and economic frontiers.[84] The Israeli agricultural frontier was typified by political-socialist time, while the urban-industrial frontier, and even more so the technological frontier, served as both incubators of and catalysts for capitalist time. This differentiation, however, has been somehow blurred by several developments. First, the many years of the strong influence of socialist notions on Israeli urban-economic life in general, and on temporality in particular, have left their imprint. The widespread acceptance of tenure as a basic value in labor relations, for example, has reduced the significance of productivity, especially in the years up to the 1970s, when newer industries changed this practice (which continues to be a burden on public services). Second, the modernization of the agricultural sector, which started in the 1960s, meant among other things the penetration of capitalist notions of time to this strongly socialist sector. The result is that increased production using less time has become a central consideration even here.

Time has continued to be a political dimension and resource. This dimension of time was strengthened after 1977, with the Likud government's attempting a speedy settlement of Judea, Samaria, and Gaza. The political movements and parties that supported this settlement campaign, especially Gush Emunim, revived the religious temporal dimension of big time. They looked on the settlement of the territories as expediting Messianic times.[85] It is interesting to compare this renewed emphasis on the big time of longue duree with Gross' observations of contemporary nation states. Gross suggests that capitalism has challenged nation-states in its ahistorical nature. The accent of capitalism is on the present and on "measured duration". Thus, he argues, "some states today do appear less inclined than formerly to base arguments for their raison d'etre upon temporal considerations. Rather, they seem more disposed to base them on strictly legal criteria, or on their ability to administer effectively, or to reproduce a more or less dependable framework for everyday life."[86]

Putting this argument into an Israeli context provides some very interesting insights. First, it is difficult to state that Israel as a nation-state has moved completely into this phase, because of both the unique role of the past and future longue duree in Judaism and the historical dimension attributed to the Arab-Israeli conflict. Second, it is even more difficult to classify the major political camps along the transition from a "temporal" nation-state to an "atemporal" capitalist state. The Likud, which advocates a free economy, emphasizes national history as a basic national value and as a major argument for keeping Judea and Samaria in Israeli hands. Labor, which established the new nation-state and its emphasis on national rather than religious motives, advocates law and order and an emphasis on the present rather than on the ancient past and far future. Labor, however, is still a proponent, though a weaker one of socialist elements in the economy.

In Chapter 2, we noted that the heavy emphasis on the quantitative-economic significance of time in capitalism amounts to a spatialization of time.[87] This phenomenon has become true of Israeli society. Here, though. the spatialization of time has additional significance -- namely, the religious spatialization of time, though in a different sense of spatialization. Big time, as Messianic time, has become strongly connected with spatial deeds, i.e., settlement activities. The spatial deeds are supposed to expedite or to be part of Messianic times. This religious conception of time has been strongly attached to space by advocates of the settlement process in Judea, Samaria, and Gaza.

The governmental use of time as a political resource for spatial action through settlement activities changed hands in 1977, from Labor to Likud, the political inheritors of Revisionist Zionism. The settlement activities of the Likud are interesting in light of the Revisionists' past rejection of settlements as not crucial for gaining sovereignty. The use of time for this purpose by Likud has been rather intensive because of both internal and external political pressures against its settlement activities. Time has thus been viewed as a political resource in the sense of achieving a Jewish presence in Judea, Samaria, and Gaza. The approach was similar to the post-1937 Labor view in pioneering Israel, though without any social goals accompanying the political aims.

The increasing importance of capitalist time and the spatialization of time do not necessarily complement earlier Judaic notions of time. The issue of "summer time," as daylight savings time is called in Israel, exemplifies a conflict between Judaic notions of time and Western capitalist ideas. For several years, energy experts, political parties, and private persons called for the institution of "daylight savings time," as is customary in North America and Western Europe. The reasoning was similar: for purposes of energy savings, increased productivity, and a higher quality of life. On the other hand, the Rabbis and the religious parties were opposed, claiming that "summer time" would make it difficult to conduct the daily morning worship before going to work. Furthermore, "summer time" would increase the violation of the Sabbath, since the end of the holy day would be delayed by one hour on Saturday nights, the favorite "night out." The issue was hotly debated in the Knesset (parliament), where the Minister of Interior, who represented a religious party, vetoed "summer time" by using a law instituted by the British Mandate. The issue later came before the supreme court, which ordered the introduction of "summer time" in the summer of 1983.

A further example of changes in temporality relates to municipal regulations prohibiting the operation of cinemas during the Sabbath (the equivalent of America's so-called blue laws). When these regulations were accepted in Israeli towns, mostly in the 1950s, they were interpreted by the religious parties as making for a Sabbath atmosphere in public life. Labor and the other political parties saw them as social regulations, controlling work and rest times or as part of a political deal with the religious parties. Recently there have been growing pressures for change, and a judge has ruled these ordinances invalid, since they, among other things, cause financial losses to cinema owners. Thus, the Sabbath was interpreted as an economic resource rather than as religious or rest time.

Spatiality in Modern Israeli Society

Space, too, has received increasing economic importance in modern Israel. Space was considered an economic dimension in pioneering Israel by middle-class Zionism. Now the view of space as an economic dimension has been made into a nation-wide attitude. The importance of space as an economic resource finds special expression at the urban level, where much land is privately owned and where urban expansion requires more land. Thus, the share of land in real estate costs increased and sharp geographical differentiations regarding land prices developed among urban zones since the 1960s. The rise of capitalism also turned space into an important production factor, so that the location of industries has been determined, among other things, by land prices. Like time, though, space has continued to serve as a political resource, too. This has been the case especially with regard to Judea, Samaria, and Gaza since 1977, but also with the Golan Heights, the Galilee, and East Jerusalem, where major settlement efforts have been taking place. It was claimed, therefore, that "the creation of the state of Israel in 1947 and the repossession of the site of the Temple of Jerusalem in 1967 have reawakened acutely the intense symbolism of place and land in Jewish consciousness."[88] The importance of space as a religious dimension and resource has increased since 1967 and even more so since 1977 among the supporters of settlements in the disputed territories. This had mainly to do with the fact that Judea and Samaria are the regions in which the ancient kingdoms of Israel and Judea had been located.

The importance of space as an economic resource related principally to small spaces, whether in the form of small plots or sites, or in the form of urban and suburban expansion. On the other hand, the importance of space as a political resource took the form of large space on the regional and national scales, which was true also for the importance of space as a religious resource. The construction of exurban communities in Judea and Samaria satisfied simultaneously the aspirations of individuals for economic prosperity and higher life quality and the political goals of government. This convergence has, obviously, been relevant only for those who agreed with government policy.

The use of space in intensive ways in urban areas, typical of pioneering Israel, has gradually changed to more extensive uses in the form of suburbanization and exurbanization. This transformation has not only been the result of continued massive urban growth, but it had also to do with changes in residential life styles: moving from small to medium and large apartments. The trend was coupled with a desire of the middle class to move into single-family dwellings. As mentioned, this change in residential values has coincided with governmental population objectives since 1977. The periphery of Jerusalem is Judea, the periphery of Tel-Aviv is Samaria, and that of Haifa is the Galilee. The political dimension of space that has called for an extensive use of space at the regional and national levels now, at least partially, coincides with the extensive use of space favored by urbanites.[89]

The increased economic significance of land has sometimes stood in contradiction to religious aspects pertaining to land. A case in point was the dispute over the ancient cemetery of Tiberias that erupted in 1984. This cemetery, inactive for centuries, consists of layers of ancient cemeteries going back some two thousand years to the Mishnaic era. The site is located in the backyard of a major hotel that started to undergo expansion. Ultra-orthodox Jewish circles demanded that work be stopped completely, while extreme secular circles called for continued normal construction on the site. A compromise was achieved, whereby part of the cemetery would remain untouched and groundwork in the remainder would proceed with extreme caution under Rabbinical supervision.

Despite the increased importance of time, it has not replaced space as the more important dimension. We noticed in Chapter 3 that a dominance of time over space is typical of capitalist societies, since time is used intensively while space is used extensively. This process we regarded as a temporalization of space, taking place side by side with the spatialization of temporality. Although time has received, in contemporary Israel, a strong "spatialized" economic significance, it is still heavily used for the achievement of spatial objectives from the political and religious viewpoints. We termed these objectives as other forms of time-spatialization. Given the still crucial, non-economic significance of space, one cannot (yet) view time as more important than space in Israeli society.

Conclusion

We have attempted to show three "layers" in Israeli spatiality and temporality: Judaic, Zionist-socialist, and capitalist. They are equivalent to the traditional, nation-state, and capitalist modes proposed by Gross for the <u>longue duree</u> of Western societies.[90] Judaic spatiality has changed much along time, temporality only very little. Thus, Judaic temporality has been interpreted as being stronger than Judaic spatiality. This interpretation is true for most aspects of temporality and spatiality, if their significance as resources is excluded. Here space seems more important than time. In socialist Zionism, space and time were firstly viewed as political dimensions and resources within a secular framework. In this context, space was shown to be more important than time. Pioneering Israel was also typified by the gradual evolution of political pluralism and the coexistence of other notions of time and space with the socialist concept. The socialist-Zionist conceptions of time and space have served, however, as a leading ideology since Labor led the Jewish population. Indeed, several of their notions of temporality and spatiality were later adopted by Likud. In contemporary, capitalist-oriented Israeli society, time and space have received economic significance as production and consumption resources. The political connotation has not disappeared, and space is still more important than time. The capture of Judea and Samaria in 1967 has led to a revival of certain elements of Judaic spatiality and temporality.

The processes of change in Israeli temporality and spatiality have immense ideological loadings that are typical of transitions in other values and structures in the society. The Judaic notions of time and space are ancient; those of Zionism, socialism and capitalism started to evolve in the nineteenth century and were imported into European Jewish society and, through immigration, into the Land of Israel. The processes of change may be summarized as follows. The weakening of Judaic spatiality and temporality stemming from secularization trends was simultaneously coupled with a strengthening of socialist-Zionist temporality and spatiality, starting in the early 1900s and culminating in the 1920s and 1930s. At the same time, the nesting of Zionist-nationalistic and capitalist notions of time and space was taking place. The dominance of Zionist-socialist time and space in pioneering Israel did not prevent a penetration of Judaic elements into socialist time-space. This blending process continued in later Israeli society, when a gradual shift occurred toward capitalist and nationalist notions of time and space; nevertheless, political elements of time-space were left strong.

The current pluralism in time-space notions in Israeli society reflects not only conflicts among the several approaches but interactive influences as well. All this leads to a minimal consensus on temporality and spatiality, consisting of a variety of elements originating in different approaches. Examples of these elements are the use of the Jewish calendar, the use of settlements as a political tool, and the increased emphasis on time as an economic resource. Israeli society may thus be typified as having both strong temporality and strong spatiality, the first, partially, of ancient Judaic origin and the second with modern Zionist and capitalist roots.

Spatiality and temporality are interrelated in each of the three major approaches described here. Within the Judaic framework, the (spatial) return to the Holy Land added to the expectation of Messianic times, and vice versa. In modern Israel, the expectations of the Messianic future led the settlement movement of Judea and Samaria. Finally, in socialist Zionism, there is a subordination of time to space; in capitalism, space-use becomes more extensive, when time-use becomes more intensive.

NOTES TO CHAPTER 5

1. The term "new society" might be interpreted to have some very specific connotations (see Elazar, 1970). Here it is used in its rather literal meaning.
2. On cultural and religious meanings of societal time, see e.g. Eliade, 1959; Goffman, 1959; Zerubavel, 1981. On the changing value of societal time under different political-economic systems see e.g., Thompson, 1967; Thrift, 1981.
3. Ilchman, 1970.
4. Gross, 1985.
5. Gregory, 1978, p. 119.
6. Giddens, 1979, p. 6.
7. Pred, 1983, p. 53.
8. E.g., Goldman, 1978; Zerubavel, 1981.
9. Goldman, 1978, p. 42.
10. Zerubavel, 1981, p. 105.
11. Goldman, 1978, p. 36.
12. Houston, 1978, p. 230.
13. Heschel, 1951, p. 10.
14. Rosenzweig, 1972, pp. 310-311.
15. Talmud, Sabbath, 118.
16. On linear time in Judaism, see e.g. Eliade, 1959, pp. 110-111. On cyclical time in Judaism see Zerubavel, 1981, pp. 112-115.
17. Eliade, op. cit.
18. The idea of "pacemakers" in time is presented by Parkes and Thrift, 1975. See also Parkes and Thrift, 1980.
19. Goldman, 1978, p. 49.
20. Exodus, 12.
21. Talmud, Berachot, 35, 2.
22. Luzzato, 1740, ch. 11.
23. Heschel, 1951, p. 8.
24. ibid., pp. 6-7.
25. Houston, 1978, p. 229.
26. Davies, 1982, pp. 8-9.
27. Bereshit (Genesis) Rabba, 1.
28. For a detailed discussion of the three holinesses, see Talmudic Encyclopedia 2, 1952, pp. 199-222.
29. Davies, 1982, p. 18.
30. On Torah as "movable territory" see Maier, 1975. On the synagogue as such see Shilhav, 1983. On temporality substituting spatiality see Zerubavel, 1981.
31. Davies, 1982, pp. 93-95.
32. For Abraham see Genesis 22, 4; for Jacob see Genesis 28, 11, and for Moses see Exodus 33, 21. Urbach (1978, p. 55n) notes that this relationship was mentioned by one, and only one, version, of Midrash Tehilim.
33. Marmorstein, 1927, pp. 92-93.

34. Urbach, 1798, p. 55.
35. Bereshit (Genesis) Rabba, 68, 8.
36. :Urbach, 1798, p. 55; Marmorstein, 1927, p. 93.
37. Schechter, 1961, p. 27n.
38. Soloveichik, 1966.
39. Steinsaltz, 1982.
40. Hoffmann, 1898, IV, p. 170, followed by Esh, 1957, p. 80.
41. Deuteronomy, 26, 2. On the high-court as "place" see Rabbi Elieser of Modin, in Mechilta, 52, b.
42. Mishna, Kelim, 1, 7-9.
43. Neusner, 1979.
44. Shilhav, 1983.
45. See Maimonides, Laws of the Temple, 2, 1-2.
46. Sce Tossefta, Avoda Zara, 5.
47. Bereshit (Genesis) Rabba, 64.
48. Shilhav, 1983.
49. Mishna, Kidushin, 36, 2; 37, 1.
50. See also Zerubavel, 1981, pp. 105-167.
51. Maimonides, Laws of Neighbors.
52. Leviticus, 25, 8-13.
53. Deuteronomy, 8.
54. Buber, 1903, p. 3; Tuan, 1978, pp. 10-11.
55. Davies, 1982, p. 126.
56. Kern, 1983, pp. 50-51. See also Heschel, 1951.
57. On the significance of the year 1957, see Segre, 1971, p. 182. On earlier signs of change see Koenig, 1952; Weinryb, 1957.
58. E.g. Davies, 1982, pp. 110-115.
59. Cohen, 1970.
60. Jabotinsky, 1915.
61. Tabenkin, 1957, p. 8.
62. ibid., 1949, p. 354.
63. Horowitz and Lissak, 1978, pp. 122-127.
64. Even-Shushan, 1972, p. 612.
65. Kimmerling, 1983, p. 20.
66. Pred, 1984.
67. See Talmud, Beitza, 25, 2.
68. Oren, 1978.
69. Akzin and Dror, 1966.
70. Tabenkin, 1949, p. 332.
71. Gonen, 1976.
72. Kimmerling, 1983, pp. 14; 21.
73. Gross, 1985, p. 55.
74. ibid., pp. 60-61.
75. Leviticus, 25, 8-13; Halevi, 1523, p. 330.
76. Horowitz and Lissak, 1978, pp. 122-127.
77. Cohen and Kliot, 1981.

78. See for example, the writings of Rabbi Isaac Kook, spiritual leader of religious Zionists, e.g. Kook, 1962, p. 214,on the one hand, and those of Rabbi Wasserman, 1969, p. 33, on the other.
79. Kimmerling, 1983, pp. 60; 152-153; 224.
80. Horowitz and Lissak, 1978, pp. 122-127.
81. Portugali, 1976.
82. Kellerman, 1986.
83. Kellerman, 1985b.
84. Elazar, 1970.
85. Sprinzak, 1985.
86. Gross, 1985, p. 77.
87. Gross, 1981; 1982.
88. Houston, 1978, p. 232.
89. See Kellerman, 1987c.
90. Gross, 1985.

CONCLUSION

We have attempted in this volume to highlight several aspects of the interrelationships of time, space, and society. Specifically the time and space of individuals and societies were compared. Then, focusing on societal time and space, we discussed the homologous relations between the two dimensions. Using the terminology and concepts developed in the first two chapters, an analysis was made of time and space within the urban context. Following this, some differences in the space and time of men and women were presented. Finally we discussed a national framework for time and space with Israel as an example.

This concluding chapter will have an eclectic nature. First, a summary of chapter conclusions will be presented. Next, using the concepts and terms proposed throughout this essay, a typology for both time and space will be proposed. This will be followed by a discussion of some of the meanings and significance of time and space. We will end with a summary of the structurationist processes of societal time and space.

Summary

Societal time is not just the aggregate time of individuals within any given societal context. This is true, too, for societal <u>versus</u> personal space. Societal time is less finite than individual time. Societal horizons extend further than those of individuals, to both past and future. This difference has a bearing on the shaping of use-norms for human time, not just on the conception of historical time. Societal space is larger than personal space, and societies act in several sites simultaneously, compared to the indivisibility of individuals.

Societal time and space serve as basic dimensions, into which society "expands" itself. Time and space, however, are also active resources used in several ways. The study of societal temporality and spatiality, or the conceptions and uses of time and space, are therefore basic to the understanding of social change. Temporality and spatiality do not necessarily have to be homologous at a certain time and place. Thus, societies may conceive of and use time and space in different modes.

Urban societies demonstrate this point, in that time has been increasingly related to as a resource in short supply, and consequently has been used intensively. Space, on the other hand, has been considered mainly a resource in ample supply; in the past, it has served as a major limiting dimension for urban growth. Space, thus, has been used extensively. The evolution and adoption of transportation and telecommunication technologies, as well as cultural and societal residential norms, have assisted the development of this contradictory trend. The increased predominance of time has imparted it the former importance of space. The specific ways of time conception have been spatialized, and space conception has been temporalized.

These typical trends in modern societies are masculine in nature.   The integration of women into production within a capitalist urban environment requires them to function in a double temporality and spatiality: regarding time, cyclical and spatial female times exist side by side with linear capitalist time; concerning space, a more restrained female spatial ability and an emphasis on small spaces have to cope with an urban world that is constantly expanding its spatial extent.

In the national framework, there has been a move from time and space determined by religion and tradition to time and space heavily influenced by the nation-state and, more recently, by the capitalist system. In Israeli society, this has meant a change of emphasis from a spiritually time-based culture to a politically spaced-based society.   Eventually, Israel will become a nation with pluralistic notions of time and space, involving elements of revived tradition and heated political significance.   These elements, moreover, are coupled with capitalist-economic notions of time and space.

A Typology for Time and Space

In order to illustrate some of the differentiations regarding both time and space, that were made in this book, we may use an extended and modified version of a typology for time proposed by Walmsley[1] (see Figures 5 and 6). Basically, both time and space are divided into those dimensions inside and those outside the body. Time inside the body consists of spatial, linear, and biological times. All these together form an individual's experiential time. Time outside the body may be divided into universal time and cultural time. Universal time is ordered and controlled by clocks and calendars. Cultural time involves three forms of time conception: time as a dimension, cyclical time, and linear time. The conception of time as a dimension calls for a passive notion of time; a linear time conception assumes that time is an active ingredient of life and a crucial movement and economic resource.   Cyclical time assumes the recurrence of events. Both time inside the body and time outside the body continuously and simultaneously shape personal and societal temporalities.

The typology for space follows similar differentiations. Space, too, is divided into space inside and space outside the body. The first consists of two spatial abilities -- namely, visualization and orientation -- that lead to subjective space. Space outside the body consists, too, of universal space and cultural space. The first, normally called geometric space, determines distances and locations. Cultural space may be conceived of and used as a passive dimension, a source of aesthetics, and an active resource through several levels of spatial size (e.g., urban, national) and intensity (e.g., extensive or intensive). Both space inside and space outside the body simultaneously and continuously shape personal and societal spatialities.

Several comments have to be made regarding these typologies of time and space. First, they are not exhaustive, and the distinctions do not always point to a complete separation between the several forms of time and space. Thus, calendars are not only astronomically determined; they also reflect cultural and

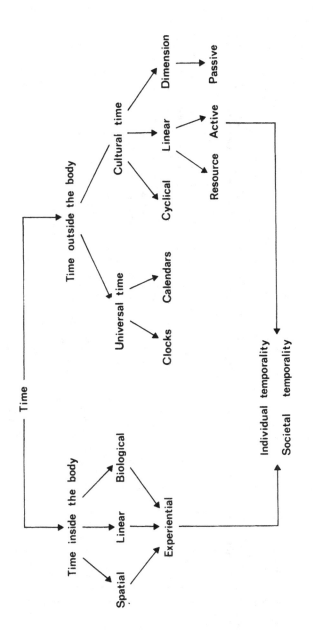

Figure 5: A Typology of Time

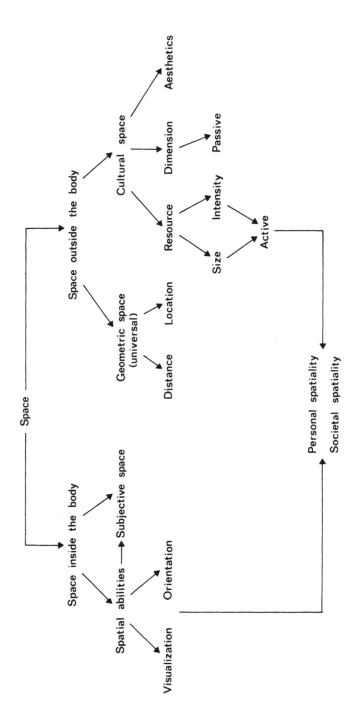

Figure 6: A Typology of Space

religious norms, as we have seen, for example, with the Hebrew calendar. By the same token, location, which was classified under geometric space, is interrelated with spatial size and intensity, which were considered to reflect cultural space. Second, the typology is endogenous for both time and space. It does not include any of the psychological, political, cultural, and economic factors that make for time and space inside and outside the body. Needless to say, the mosaic of exogenous factors of time and space is extremely wide and complex and does not lend itself to a simple presentation. Third, spatialization and temporalization are purposely excluded from the two models, since they are not processes that must occur. While temporality and spatiality are human and social aspects that are indispensible to our life, spatialization and temporalization may or may not develop, given certain cultural circumstances. Fourth, the several components and meanings of both time and space may complement or conflict with one another. This is also true of comparisons of time and space. Examples of such conflicts are female spatial-cyclical time and capitalist linear time; linear and cyclical societal notions of time; passive space and space as a resource, and female spatial needs compared to extensive capitalist spaces. On the other hand, inner linear temporality complements societal linear temporality; the active use of time fits the notion of time as a resource. By the same token, an active notion of space would fit the conception of space as a resource.

Comments on the Meanings and Significance of Time and Space

A major thesis argued throughout this book has been that the importance and significance of time and space have changed in several respects and levels in modern society. One of the more striking transitions relates to the relationship between historical-cultural time, on the one hand, and human-economic time, on the other. In the past, historical time was considered linear, and human time cyclical. Currently the opposite is true. Economic time is linear and cultural-historical time is conceived of as cyclical; or better to say, as spatial. This is the spatialization process referred to by Gross; or "the vital present," as it was termed by Cottle.[2] The process seems to encompass both personal and societal lives. It has given time the former importance of space, so that time has become the more important dimension of the two, though in a spatial way. The new capitalist mode of temporality may come into conflict with gender, as well as with the religious and political significance of time, and not only with its cultural meaning. Thus, time and space may mean different things and may be used in different ways by several groups within one society. Obviously, time and space might be interpreted differently by distinct national societies. The heavy influence of capitalism has caused its linear and spatialized modes of temporality and its extensive and temporalized spatiality to compete with other forms of temporality and spatiality.

An argument put forward in Chapter 3 related modern, intensified temporality to modern, extensive spatiality through transportation and telecommunications. The possible connections between modern temporality and spatiality deserve a second look, at least in the form of question raising, after the discussions on women and Israel, respectively in Chapters 4 and 5. It is intriguing to find out, at both the psychological and sociological-cultural levels, whether the increased intensity of time-use necessitates a more extensive space-use for individuals. This more extensive spatiality would be expressed in the form of larger homes, weekend travel, second homes, etc. The question needs both a psychological and cultural study, since, on the one hand, one may argue that an extensive spatiality is a basic human need to compensate for a more intensified temporality. On the other hand, there might be other ways to relieve pressures related to more intensive temporalities. Judaism, for example, rather than trading space for time, provides "time for time." Thus the weekly Sabbath and its mandatory atmosphere of relaxation should relieve weekday temporality intensities. Even more ineresting should be the urban spatially intensive Orient, in which temporality is no less intensive than in the West. Thus, much more light must be shed on the relationship between time and space, and between both and individuals and societies.

Structuration and Societal Time and Space

The definition of spatiality and temporality as processes through which the social becomes spatial or temporal, and vice versa, is structurationist. This is also true of the assertion that the study of the changing conceptions and uses of time and space implies that concepts of time and space are not found within time and space, but in Man and society (Chapter 2). It is, thus, our belief that time and space cannot be separated from social studies. Furthermore, even when "bracketing" is used, so that the focus is on societal spatiality and temporality, those of the individual cannot be ignored. This is due to the constant two-way interaction between these two levels. Thus, it is human action that changes societal time and space uses and conceptions; and it is also, and at the same time, societal spatialities and temporalities that shape individual spatialities and temporalities.

Societal temporality and spatiality may change, also, through technological and ideological transitions. These changes may then bring about new relationships between time and space and, eventually, transform their societal importance.

Spatiality and temporality may, thus, be described as participants in several constant and simultaneous dialectics. On the one hand, there is a structuration of time and space, so that societal spatiality and temporality influence individual spatiality and temporality. At the same time, however, individuals' spatialities and temporalities reshape those of society. Societal temporality and spatiality are also involved in a mutual two-way interaciton, so that notions of space shape notions of time, and vice versa. These changes may take place much longer, and they do not have to be in both directions at the same time. Thus, as we have seen, temporality may influence spatiality within an urban capitalist system, but not vice versa. A third process of structuration takes place between time and space, on the one hand, and social structures, on the other. This vast area of interaction has only been hinted at here and there in this volume, and it still awaits exposition.

NOTES TO CONCLUSION

1. Walmsley, 1979, P. 224.
2. Gross, 1981; Cottle, 1976.

REFERENCES

Ad Hoc Committee on Geography, Earth Sciences Division. 1965. The Science of Geography. Washington, D.C.: National Academy of Sciences - National Research Council.

Akzin, B., & Dror, Y., 1966. Israel: High Pressure Planning. Syracuse, N.Y.: Syracuse University Press.

Alexander, S., 1920. Space, Time and Deity. Book 1: Space-time. London: Macmillan (1966 edition).

Althusser, L. & Balibar, E. 1970. Reading Capital. London: New Left Books.

Appleyard, D. 1970. Styles and methods of structuring the city. Environment and Behavior, 2, 100-117.

Baker, A.R.H. 1981. A historico-geographical perspective on time and space, and on period and place. Progress in Human Geography, 5, 439-443.

Becker, G.S. 1965. A theory of the allocation of time. Economic Journal, 75, 493-517. Bell, D. 1973. The Coming of Post-Industrial Society. New York: Basic Books.

Bereshit (Genesis) Rabba. 1961. Trans. H. Freeman and I. Epstein. London: The Soncino Press.

Bergson, H. 1959. Time and Free Will: An Essay on the Immediate Data of Consciousness. Trans. F.L. Pogson, London: George, Allen & Unwin.

Berry, B.J.L. 1964. Approaches to regional analysis: A synthesis. Annals of the Association of American Geographers, 54, 2-11.

Bible - The Holy Scriptures of the Old Testament. 1966. London: British and Foreign Bible Society.

Boserup, E. 1970. Woman's Role in Economic Development. New York: St. Martin's Press.

Braudel, F. 1980. On History. Chicago: The University of Chicago Press.

Buber, M. 1903. Juedische Kuentsler. Berlin: Juedische Verlag.

Carlstein, T. 1978. Innovation, time allocation and time-space packing. In: T. Carlstein, D. Parkes, & N. Thrift (eds.), Timing Space and Spacing Time 2. Human Activity and Time Geography. London: Arnold, pp. 141-61.

Carlstein, T. 1981. The sociology of structuration in time and space: A time-geographic assessment of Giddens's theory. Svensk Geografisk Arsbok, 57, 41-57.

Carlstein, T. 1982. Time Resources, Society and Ecology: Vol. 1 Preindustrial Societies. London: George Allen & Unwin.

Carlstein, T., Parkes, D., & Thrift, N. (eds.), 1978. Timing Space and Spacing Time 3. Time and Regional Dynamics. London: Arnold.

Carnap, R. 1958. Introduction to Symbolic Logic and Its Applications. New York: Dover.

Castells, M. 1977. The Urban Question: A Marxist Approach. London: Arnold.

Chabaud, D. & Fougeyrollas, D. 1978. Travail domestique et espace-temps des femmes. International Journal of Urban and Regional Research, 2, 421-431.

Chapin, F.S. Jr. 1974. Human Activity Patterns in the City: Things People Do in Time and Space. New York: John Wiley.

Clark, C. 1951. Urban population densities. Journal of the Royal Statistical Society, 114A, 490-496.

Cohen, E. 1970. The city in the Zionist Ideology. Jerusalem Urban Studies 1. Jerusalem: Institute of Urban and Regional Studies, The Hebrew University.

Cohen, S.B. & Kliot, N. 1981. Israel's place names as reflection of continuity and change in nation-building. Names, 29, 227-248.

Cottle, T.J. 1976. Perceiving Time: A Psychological Investigation with Men and Women. New York: John Wiley.

Crevecoeur, J.H. St. John de. 1782. Letters from an American Farmer. London: J.M. Dent.

Curry, L. 1978. Position, flow and person in theoretical economic geography. In: T. Carlstein, D. Parkes, & N. Thrift (eds.), Timing Space and Spacing Time 3. Time and Regional Dynamics. London: Arnold, pp. 35-50.

Davies, W.D. 1982. The Territorial Dimension of Judaism. Berkeley, CA: University of California Press.

Dear, M.J. & Moos, A.I. 1986. Structuration theory in urban analysis: 2. Empirical application. Environment and Planning A, 18, 351-373.

de Grazia, S. 1962. Of Time, Work and Leisure. New York: Twentieth Century Fund.

Diesing, P. 1962. Reason in Society. Urbana: University of Illinois Press.

Douglas, H.P. 1925. The Suburban Trend. New York: The Century Co. Reprinted 1970. New York: Arno Press and The New York Times.

Durkheim, E. 1915. The Elementary Forms of the Religious Life. London: George Allen & Unwin.

Elazar, D.J. 1970. Israel: From Ideological to Territorial Democracy. New York: General Learning Press.

Ellegard, K., Hagerstrand, T., & Lenntorp, B. 1977. Activity organization and the generation of daily travel: Two future alternatives. Economic Geography, 53, 126-152.

Eliade, M. 1954. Cosmos and History: The Myth of the Eternal Return. New York: Harper and Row.

Eliade, M. 1959. The Sacred and the Profane. New York: Harcourt, Brace & Co.

Ericksen, J. 1977. An analysis of the journey to work for women. Social Problems, 24, 428-435.

Erickson, E.H. 1964. Inner and outer space: reflections on womanhood. Daedalus, 93, 582-606.

Erickson, E.H. 1968. Identity, Youth and Crisis. New York: Norton.

Esh, S. 1957. Der Heilige (Er Sei Gepriesen). Leiden: E.J. Brill.

Even Shushan, A. 1972. The Concentrated Hebrew Dictionary. Jerusalem: Qiryat Sefer. (Hebrew).

Everitt, J. & Cadwallader, M. 1972. The home area concept in urban analysis: the use of cognitive mapping and computer procedures as methodological tools. In: W.J. Mitchell (ed.), Environmental Design: Research and Practice. Los Angeles: University of California.

Falk, T. & Abler, R. 1980. Intercommunications, distance and geographical theory. Geografiska Annaler, 62, 59-67.

Fava, S.F. 1980. Women's place in the new suburbia. In: G.R. Wekerle et al. (eds.), New Space for Women. Boulder CO: Westview Press.

Firestone, S. 1970. The Dialectic of Sex. New York: William Morrow.

Flink, J.J. 1975. The Car Culture. Cambridge, MA: MIT Press.

Forer, P. 1978. Time-space and area in the city of the plains. In: T. Carlstein, D. Parkes, & N. Thrift (eds.), Timing Space and Spacing Time 1. Making Sense of Time. London: Arnold, pp. 99-118.

Fraser, J.T. 1968. The study of time. In: J.T. Fraser (ed.), The Voices of Time. Harmondsworth: Penguin, pp. 590-591.

Fraser, J.T. 1975. Of Time, Passion and Knowledge. New York: Braziller.

Georgescu-Roegen, N. 1971. The Entropy Law and the Economic Process. Cambridge, MA: Harvard University Press.

Ghez, G.R., & Becker, G.S. 1975. The Allocation of Time and Goods over the Life cycle. New York: Columbia University Press.

Giddens, A. 1979. Central Problems in Social Theory. Berkeley, CA: University of California Press.

Giddens, A. 1981. A Contemporary Critique of Historical Materialism. Berkeley, CA: University of California Press.

Giddens, A. 1982. Profiles and Critiques in Social Theory. Berkeley, CA: University of California Press.

Giddens, A. 1984. The Constitution of Society: Outline of the Theory of Structuration. Berkeley, CA: University of California Press.

Gilmartin, P.P. & Patton, J.C. 1984. Comparing the sexes on spatial abilities: map-use skills. Annals of the Association of American Geographers, 74, 605-619.

Goffman, E. 1959. The Presentation of Self in Everyday Life. Garden City, NJ: Doubleday.

Gold, J.R. 1980. An Introduction to Behavioural Geography. Oxford: Oxford University Press.

Goldman, E. 1978. An inquiry into the philosophy of time and its meaning to the study of Judaism. Unpublished Ed.D. dissertation, Boston University.

Gonen, A. 1976. The suburban mosaic in Israel. In: D. Amiran & Y. Ben Aryeh (eds.), Geography in Israel. Jerusalem: The Israel National Committee, International Geographical Union, 163-186.

Goody, J. 1968. Time: social organization. In: International Encyclopedia of the Social Sciences, Vol. 16. New York: Macmillan, pp. 30-42.

Gottmann, J. 1952. Le Politique des Etats et Leur Geographie. Paris: Colin.

Gould, P. 1981. Space and rum: An English note on espacien and rumian meaning. Geografiska Annaler, 63B, 1-3.

Gregory, D. 1978. Ideology, Science and Human Geography. London: Hutchinson.

Gregory, D. 1982a. Regional Transformation and Industrial Revolution: A Geography of the Yorkshire Woolen Industry. London: Macmillan.

Gregory, D. 1982b. Solid geometry: notes on the recovery of spatial structure. In: P. Gould & G. Olsson (eds.), A Search for Common Ground. London: Pion, pp. 187-219.

Gregory, D. & Urry, J. (eds.) 1985. Social Relations and Spatial Structures. New York: St. Martin's.

Gregson, N. 1986. On duality and dualism: the case of structuration and time geography. Progress in Human Geography, 10, 184-205.

Gregson, N. 1987. Structuration theory: some thoughts on the possibilities for empirical research. Environment and Planning D: Society and Space, 5, 73-91.

Gross, D. 1981. Space, time and modern culture. Telos, 50, 59-78.

Gross, D. 1982. Time-space relations in Giddens' social theory. Theory, Culture and Society, 1, 83-88.

Gross, D. 1985. Temporality and the modern state. Theory and Society, 14, 53-82.

Gurvitch, G. 1964. The Spectrum of Social Time. Dordrecht: Reidel.

Hagerstrand, T. 1970. What about people in regional science? Papers and Proceedings of the Regional Science Association 24, 7-21.

Hagerstrand, T. 1973. The domain of human geography. In: R.J. Chorley (ed.), Directions in Geography. London: Methuen, pp. 67-87.

Hagerstrand, T. 1975. Space, time and human conditions. In: A. Karlqvist, L. Lundqvist, & F. Snickers (eds.), Allocation of Urban Space. Farnborough: Saxon House, pp. 3-12.

Halevi, A. 1969. Book of Education (Sefer Hahinuch). Jerusalem: Eshkol. Originally published in 1523 (Hebrew).

Hall, E.T. 1966. The Hidden Dimension. New York: Doubleday.

Hall, P. & Hay, D. 1980. Growth Centers in the European Urban System. Berkeley, CA: University of California Press.

Harris, J.L. 1978. Sex differences in spatial abilities: possible environmental, genetic, and neurological factors. In: M. Kinsbourne, (ed.), Asymmetrical Functions of the Brain, New York: Cambridge University Press, pp. 405-522.

Hart, R. 1979. Children's Experience of Place. New York: Irvington.

Hartmann, H. 1976. Capitalism, patriarchy, and job segregation by sex. Signs, 1, 137-169.

Hartshorn, R. 1950. The functional approach in political geography. Annals of the Association of American Geographers, 40, 985-103.

Harvey, D. 1969. Explanation in Geography. New York: St. Martin's Press.

Harvey, D. 1973. Social Justice and the City. London: Arnold.

Harvey, D. 1984. On the history and present condition of geography: an historical materialist manifesto. The Professional Geographer, 36, 1-11.

Hayden, D. 1984. Redesigning the American Dream. New York: Norton.

Heller, A. 1978. Renaissance Man. London: Routledge & Kegan Paul.

Heschel, A.J. 1951. The Sabbath. New York: Farrar, Strauss & Giroux.

Hoffman, D.Z. 1898. Mishnaiot. Die sechs Ordnungen der Mischna. Berlin: H. Itzkowski. Reprinted: Frankfurt and Wiessbaden, 1924.

Holly, B.P. 1978. The problem of scale in time-space research. In: T. Carlstein, D. Parkes, & N. Thrift (eds.), Timing Space and Spacing Time 3. Time and Regional Dynamics. London: Arnold, pp. 5-18.

Hopkins, L.B. 1980. Inner space and outer space identity in contemporary females. Psychiatry, 43, 1-12.

Horowitz, D. & Lissak, M. 1978. Origins of Israeli Polity: Palestine under the Mandate. Trans. C. Hoffman. Chicago: University of Chicago Press.

Horton, F.E. & Reynolds, D.R. 1971. Action space differentials in cities. In: H. McConell & D. Yaseen (eds.), Perspectives in Geography: Models of Spatial Interaction. Dekalb, IL: Northern Illinois University Press, pp. 83-102.

Houston, J.M. 1978. The concepts of 'place' and 'land' in the Judaeo- Christian tradition. In: D. Ley & M.S. Samuels (eds.), Humanistic Geography. Chicago: Maaroufa Press, pp. 224-237.

Ilchman, W.F. 1970. New time in old clocks: productivity, development and comparative public adnministration. In D. Waldo (ed.), Temporal Dimensions of Development Administration. Durham, NC: Duke University Press, pp. 135-178.

Innis, H.A. 1951. The Bias of Communication. Toronto: University of Toronto Press.

Jabotinsky, Z. 1915. Back to the charter. Die Triebune (November 15) (Yiddish).

Jackle, J.A., Brunn, S. & Roseman, C.C. 1976. Human Spatial Behavior. North Scituate, MA: Duxbury Press.

Jackson, K.T. 1985. Crabgrass Frontier. New York: Oxford University Press.

Janelle, D.G. 1969. Spatial reorganization: a model and concept. Annals of the Association of American Geographers, 58, 348-364.

Kant, I. 1961. Critique of Pure Reason. Trans. N.K. Smith. London: Macmillan.

Katznelson, I. 1979. Community, capitalist development and emergence of class. Politics and Society, 9, 203-37.

Kellerman, A. 1981. Time-space approaches and regional study. Tijdschrift voor Economische en Sociale Geografie, 72, 17-27.

Kellerman, A. 1983. The suburbanizaton of retail trade: the Israeli case. Area, 15, 219-22.

Kellerman, A. 1984. Telecommunications and the geography of metropolitan areas. Progress in Human Geography, 8, 222-246.

Kellerman, A. 1985a. The suburbanization of retail trade: a U.S. nationwide view. Geoforum, 16, 15-23.

Kellerman, A. 1985b. Population dispersal: forecasting and reality in the four million population plan for Israel. Geography Research Forum, 8, 53-72.

Kellerman, A. 1985c. The evolution of service economies: a geographical perspective. The Professional Geographer, 37, 133-143.

Kellerman, A. 1986. Characteristics and trends in the Israeli service economy. The Service Industries Journal, 6, 205-206.

Kellerman, A. 1987a. Structuration theory and attempts at integration in human geography. The Professional Geographer, 39, 267-274.

Kellerman, A. 1987b. Time-space homology: A societal-geographical perspective. Tijdschrift voor Economische en Sociale Geografie, 78, 251-264.

Kellerman, A. 1987c. 'To Become a Free Nation in Our Land': Transitions in the Priorities of Zionist Objectives and Their Geographical Implementation. Monogeography 5. Haifa: University of Haifa, Department of Geography (Hebrew).

Kellerman, A. 1988. The integrity of integration. The Professional Geographer, 40, 221-222.

Kellerman, A. & Krakover, S. 1986. Multi-sectoral urban growth in space and time: an empirical approach. Regional Studies, 20, 117-129.

Kern, S. 1983. The Culture of Time and Space 1880-1918. Cambridge, MA: Harvard University Press.

Kimmerling, B. 1983. Zionism and Territory, Berkeley, CA: University of California, Institute of International Studies, Research Series No. 51.

Knight, D.B. 1982. Identity and territory: Geographical perspectives on nationalism and regionalism. Annals of the Association of American Geographers, 72, 514-531.

Koenig, S. 1952. Immigration and culture conflict in Israel. Social Forces, 31, 144-148.

Kolaja, J. 1969. Social System and Time and Space. Pittsburgh: Duquesne University Press.

Kook, A.Y. 1962. Letters (Igrot Hariyah) Jerusalem: Rabbi Kook Institute (Hebrew).

La Gory, M. & Pipkin, J. 1981. Urban Social Space. Belmont, CA: Wadsworth.

Lakoff, G. & Johnson, M. 1980. Metaphors We Live By. Chicago: University of Chicago Press.

Landes, D.S. 1983. Revolution in Time. Cambridge, MA: Harvard University Press.

Lefebvre, H. 1970. La Revolution Urbaine. Paris: Gallimard.

Lefebvre, H. 1974. La Production de l'Espace. Paris: Anthropos.

Lefebvre, H. 1976. The Survival of Capitalism. London: Allison & Busby.

Lefebvre, H. 1977. Reflections on the politics of space. In: R. Peet (ed.), Radical Geography. Chicago: Maaroufa Press, pp. 339-352.

Ley, D. 1983. A Social Geography of the City. New York: Harper & Row.

Linder, S.B. 1970. The Harried Leisure Class. New York: Columbia University Press.

Lipietz, A. 1975. Structuration de l'espace probleme foncier et amenagement du territoire. Environment and Planning A, 7, 415-425.

Lowe, D.M. 1982. History of Bourgeois Perception. Chicago: University of Chicago Press.

Loyd, B. 1975. Woman's place, man's place. Landscape, 20, 10-3.

Luhmann, N. 1982. The Differentiation of Society. Trans. S. Holmes, and C. Larmore. N.Y.: Columbia University Press.

Luzzatto, M.A. 1966. Messilat Yesharim. Trans. S. Silverstein. Jerusalem: Boys Town. Originally published in 1740.

Lynch, K. 1960. The Image of the City. Cambridge, MA: The MIT Press.
Lynch, K. 1972. What is This Place? Cambridge, MA: The MIT Press.
Lynch, K. 1976. Managing the Sense of a Region. Cambridge, MA: The MIT Press.
Maccoby, E.E. & Jacklin, C.N. 1974. The Psychology of Sex Differences. Stanford, CA: Stanford University Press.
McDowell, L. 1983. Towards an understanding of the gender division of urban space. Environment and Planning D: Society and Space, 1, 59-72.
McGee, M.G. 1979. Human Spatial Abilities: Sources of Sex Differences. New York: Praeger.
Mackensie, S. & Rose, D. 1983. Industrial change, the domestic economy and home life. In: J. Anderson, S. Duncan, & R. Hudson, R. (eds.), Redundant Spaces in Cities and Regions. IBG Special Publication 15, London: Academic Press.
Maier, E. 1975. Torah as moveable territory. Annals of the Association of American Geographers, 65, 18-23.
Maimonides, (Moses Ben Maimon). 1955. The Code of Maimonides. New Haven, CT: Yale University Press.
Markusen, A.R. 1981. City spatial structure, women's household work, and national urban policy. In: C.R. Stimpson, et al. (eds.), Women and the American City. Chicago: University of Chicago Press.
Marmorstein, A. 1927. The Old Rabbinic Doctrine of God. London: Oxford University Press.
Matthews, M.H. 1984. Cognitive mapping abilities of young boys and girls. Geography, 69, 327-336.
Matthews, M.H. 1987. Gender home range and environmental cognition. Transactions of the British Institute of Geographers, 12, 43-56.
Mazey, M.E. & Lee, D.R. 1983. Her Space, Her Place: A Geography of Women. Washington, D.C.: The Association of American Geographers.
Mechilta Derabbi Yishmael. 1949. Trans. J.Z. Lauterbach. Philadelphia: Jewish Publication Society of America.
Meinig, D.W. 1979. The beholding eye. In: D.W. Meinig (ed.), The Interpretation of Ordinary Landscapes. New York: Oxford University Press, pp. 33-50.
Melbin, M. 1978a. Night as frontier. American Sociological Review, 43, 3-22.
Melbin, M. 1978b. The colonization of time. In: T. Carlstein, D. Parkes, & N. Thrift (eds.), Timing Space and Spacing Time 2. London: Arnold, pp. 100-113.
Miller, R. 1983. The Hoover in the garden: middle class women and suburbanization, 1850-1920. Environment and Planning D: Society and Space, 1, 73-87.
Mishna Kelim and Kiddushin. 1938. Trans. H. Danby. London: H. Milford.
Moore, W.E. 1963. Man, Time and Society. New York: John Wiley.
Moos, A.I. & Dear, M.J. 1986. Structuration theory in urban analysis: 1. theoretical exegesis. Environment and Planning A, 18, 231-252.
Morrill, R.L. 1984. Recollections of the 'quantitative revolution's' early years: The University of Washington 1955-65. In: M. Billings, D. Gregory, & R.

Martin (eds.), Recollections of a Revolution: Geography as a Spatial Science. London: Macmillan, pp. 57-72.

Muller, P.O. 1981. Contemporary Suburban America. Englewood Cliffs, NJ: Prentice-Hall.

Mumford, L. 1934. Technics and Civilization. New York: Harcourt, Brace.

Mumford, L. 1961. The City in History. New York: Harcourt, Brace.

Munroe, R.L. & Munroe, R.H. 1971. Effect of environmental experience on spatial ability in East African society. Journal of Social Psychology, 83, 15-22.

Naisbitt, J. 1984. Megatrends. New York: Warner.

Neusner, J. 1979. Map without territory: Mishnah's system of sacrifice and sanctuary. History of Religions, 19, 103-127.

Nyborg, H. 1983. Spatial ability in men and women - review and new theory. Advances in Behaviour Research and Therapy, 5, 89-140.

Oakley, A. 1974. Woman's Work: The Housewife, Past and Present. New York: Vintage Books.

Oren, E. 1978. Settlement During the Years of Struggle. Jerusalem: Yad Ben-Zvi. (Hebrew).

Orleans, P.A. & Schmidt, S. 1972. Mapping the city: environmental cognition of urban residents. In: W.J. Mitchell, (ed.), Environmental Design: Research and Practice. Los Angeles: University of California Press.

Orme, J.E. 1969. Time, Experience and Behaviour. London: Iliffe Books.

Palm, R. & Pred, A. 1974. A time-geographic perspective on problems of inequality for women. Berkeley: University of California: Institute of Urban and Regional Development, Working Paper 236.

Parker, S.R. & Smith, M. 1976. Work and leisure. In: R. Dubin (ed.), Handbook of Work, Organization and Society. Chicago: Rand McNally, pp. 37-62.

Parkes, D.N. 1971. Urban clocks? Section 21, Australia and New Zealand Association for the Advancement of Science. University of Queensland. (Mimeo).

Parkes, D.N. 1973. Timing the city: a theme for urban environmental planning. Royal Australian Planning Institute Journal, 12, 130-135.

Parkes, D.N. 1974,. Themes on time in urban social space: an Australian study. Seminar Series, 26, Department of Geography, University of Newcastle-upon-Tyne, Newcastle, England.

Parkes, D.N. & Thrift, N. 1975. Timing space and spacing time. Environment and Planning A, 7, 651-670.

Parkes, D.N. & Thrift, N.J. 1980. Times, Spaces and Places: A Chronogeographic Perspective. London: John Wiley.

Pettifer, J. & Turner, N. 1984. Automania. London: Collins.

Piaget, J. 1971. Genetic Epistemology. New York: Norton.

Pocock, D.C.D. 1976. Some characteristics of mental maps: an empirical study. Transactions of the Institute of British Geographers, 1, 493-512.

Porteous, J.D. 1977. Environment and Behavior. Reading, MA: Addison-Wesley.

Portugali, J. 1976. The effect of nationalism on the settlement pattern of Israel. Unpublished Ph.D. dissertation, London School of Economics.

Pratt, J.H. 1984. Home teleworking: a study of its pioneers. Technological Forecasting and Social Change, 25, 1-14.

Pred, A. 1977a. Editorial note. Economic Geography, 53, 98.

Pred, A. 1977b. The choreography of existence: comments on Hagerstrand's time-geography and its usefulness. Economic Geography, 53, 207-21.

Pred, A. 1978. The impact of technological and institutional innovations on life content: some time-geographic observations. Geographical Analysis, 10, 345-72.

Pred, A. 1982. Social reproduction and the time geography of everyday life. In: P. Gould & G. Olsson (eds.), A Search for Common Ground. London: Pion, pp. 157-186.

Pred, A. 1983. Structuration and place: on the becoming of sense of place and structure of feeling. Journal of the Theory of Social Behavior, 13, 45-68.

Pred, A. 1984. Place as historically contingent process: structuration and the time-geography of becoming place. Annals of the Association of American Geographers, 74, 279-297.

Relph, E.C. 1976. Place and Placelessness. London: Pion.

Rich, O.J. 1973. Temporal and spatial experience as reflected in the verbalizations of multiparous women during labor. Maternal-Child Nursing Journal, 2, 239-325.

Rose, C. 1977. Reflections on the notion of time incorporated in Hagerstrand's time geographic model of society. Tijdschrift voor Economische en Sociale Geografie, 68, 43-9.

Rose, C. 1987. The problem of reference and geographic structuration. Environment and Planning D: Society and Space, 5, 93-106.

Rosenzweig, F. 1972. The Star of Redemption. Trans. W. Hallo. Boston: Beacon Press.

Rothblatt, D.N., Garr, D.J., & Sprague, J. 1979. The Suburban Environment and Women. New York: Praeger.

Saegert, S. 1983. Untitled comments. In: D. Amedeo, J.B. Griffin, & J.J. Potter (eds.), EDRA 1983, p. 186.

Saegert, S. 1984. Masculine cities and feminine suburbs: polarized ideas, contradictory realities. In: C.R. Stimpson, et al. (eds.), Women and the American City. Chicago: University of Chicago Press.

Saegert, S. & Hart, R. 1978. The development of environmental competence in girls and boys. In: M.Salter, (ed.), Play: Anthropological Perspectives. New York: Leisure Press, pp. 157-176.

Salomon, I. & Salomon, M. 1984. Telecommuting: the employee's perspective. Technological Forecasting and Social Change, 25, 15-28.

Sartre, J.P. 1960. Critique de la Raison Dialectique. Paris: Gallimard.

Saunders, P. 1986. Social Theory and the Urban Question. 2nd ed. London: Hutchinson.

Schechter, S. 1961. Aspects of Rabbinic Theology. New York: Schocken Books.

Schwartz, B. 1978. The social ecology of time barriers. Social Forces, 56, 1203-1220.

Schwartz-Cowan, R. 1983. More Work for Mother. New York: Basic Books.
Segre, V.D. 1971. Israel: A Society in Transition. London: Oxford University Press.
Shackle, G.L.S. 1978. Time, choice and uncertainty. In: T. Carlstein, D. Parkes, & N. Thrift (eds.), Timing Space and Spacing Time 1. Making Sense of Time. London: Arnold, pp. 47-55.
Shahar, S. 1983. The Fourth Order: A History of Women in the Middle Ages. Tel Aviv: Dvir (Hebrew).
Shamai, S. & Kellerman, A. 1985. Conceptual and experimental aspects of regional awareness: an Israeli case study. Tijdschrift voor Economische en Sociale Geografie, 76, 88-99.
Sherman, J.A. 1978. Sex Related Cognitive Differences: An Essay on Theory and Evidence. Springfield IL: Charles C. Thomas.
Shilhav, Y. 1983. Principles for the location of synagogues: symbolism and functionalism in a spatial context. The Professional Geographer, 35, 324-329.
Soja, E.W. 1979. Between geographical materialism and spatial fetishism: some observations on the development of Marxist spatial analysis. Antipode, 11, 3-11.
Soja, E.W. 1980. The socio-spatial dialectic. Annals of the Association of American Geographers, 70, 207-225.
Soja, E.W., 1982. Spatiality, politics and the role of the state. Latin American Regional Conference, 2 (IGU, Rio de Janeiro), 245-253.
Soja, E.W. 1985. Regions in context: spatiality, periodicity, and the historical geography of the regional question. Environment and Planning D: Society and Space, 3, 175-190.
Soloveichik, J.B. 1966. Sacred and profane: Kodesh and Chol in world perspective. Gesher, 3(1), 5-29.
Soule, G.H. 1955. Time for Living. New York: Viking Press.
Spencer, C.P. & Weetman, M. 1981. The microgenesis of cognitive maps: a longitudinal study of new residents of an urban area. Transactions of the Institute of British Geographers, 6, 375-384.
Sprinzak, E. 1985. The iceberg model of political extremism. In: D. Newman, (ed.), The Impact of Gush Emunim. London: Croom Helm, pp. 27-45.
Steinsaltz, A. 1982. The Passover Haggadah. Jerusalem: Carta (Hebrew).
Stimpson, C.R., Dixler, E., Nelson, M.J. & Yatrakis, K.B. 1981. Foreward. In: C.R. Stimpson, et al. (eds.), Women and the American City. Chicago: University of Chicago Press.
Storper, M. 1985. The spatial and temporal constitution of social action: a critical reading of Giddens. Environment and Planning D: Society and Space, 3, 407-424.
Tabenkin, Y. 1949. Speeches. Vol. 4. Tel Aviv: Hakibbutz Hameuchad (Hebrew).
Tabenkin, Y., 1957. Sources for the Counsellor. Vol. 1. Tel Aviv: Hakibbutz Hameuchad (Hebrew).
Talmud, Beitza, Berachot and Sabbath, 1962-1983. London: The Soncino Press.
Talmudic Encyclopedia. 1952. Jerusalem: Talmudic Encyclopedia (Hebrew).

Taylor, P.J. & Parkes, D.N. 1975. A Kantian view of the city: a factorial ecology experiment in space and time. Environment and Planning A, 7, 671-88.

Thompson, E.P. 1967. Time, work-discipline and industrial capitalism. Past and Present, 38, 56-97.

Thrift, N. 1977a. Time and theory in human geography. Part I. Progress in Human Geography, 1, 65-101.

Thrift, N. 1977b. Time and theory in human geography. Part II. Progress in Human Geography, 1, 413-457.

Thrift, N. 1981. Owner's time and own time: the making of a capitalist time consciousness, 1300-1880. Space and Time in Geography, Lund Studies in Geography, B48, 56-84.

Thrift, N. 1983. On the determination of social action in space and time. Environment and Planning D: Society and Space, 1, 23-57.

Thrift, N. 1985. Bear and mouse or bear and tree? Anthony Giddens' reconstitution of social theory. Sociology 19, 609-623.

Toffler, A. 1980. The Third Wave. London: Pan.

Tossefta, Avoda Zara. 1981. Trans. J. Neusner. New York: Ktav.

Tuan, Y.F. 1974. Topophilia. Englewood Cliffs, NJ: Prentice-Hall.

Tuan, Y.F. 1978. Space, time, place: a humanistic frame. In: T. Carlstein, D. Parkes & N. Thrift (eds.), Timing Space and Spacing Time 1: Making Sense of Time. London: Arnold, pp. 7-16.

Ullman, E. 1974. Space and/or time: opportunity for substitution and prediction. Transactions of the Institute of British Geographers, 63, 135-139.

Urbach, E. 1978. The Sages: Chapters of Beliefs and Opinions. Jerusalem: Magnes Press and the Hebrew University (Hebrew).

Urry, J. 1981. Localities, regions and social class. International Journal of Urban and Regional Research, 5, 455-473.

Urry, J. 1985. Social relations, space and time. In: D. Gregory & J. Urry (eds), Social Relations and Spatial Structures. London: Macmillan, pp. 20-48.

Van den Berg, L., Drewett, R., Klaassen, L.H., Rossi, A. & Vijverberg, C.H.T. 1982. A Study of Growth and Decline. Urban Europe, Vol. 1. Oxford: Pergamon Press.

Walmsley, D.J. 1979. Time and human geography. Australian Geographical Studies, 17, 223-229.

Warntz, W. 1968. Global science and the tyranny of space. Papers of the Regional Science Association, 20, 7-19.

Wasserman, E. 1969. The Trace of Messiah (Ikvta Demeshicha). Jerusalem (Hebrew).

Webber, M.J. 1982. Location of manufacturing activity in cities. Urban Geography, 3, 203-223.

Webber, M.M. 1963. Order in diversity: community without propinquity. In: L. Wingo, Jr. (ed.), Cities and Space: The Future Use of Urban Land. Baltimore: Johns Hopkins University Press, pp. 23-54.

Wekerle, G.R. 1981. Women in the urban environment. In: C.R. Stimpson, et al. (eds.), Women and the American City. Chicago: University of Chicago Press.

Wekerle, G.R., Peterson, R., & Morley, D. 1980. Introduction. In: G.R. Wekerle et al. (eds.), New Space for Women. Boulder: Westview Press.

Weinryb, B.D. 1957. The impact of urbanization in Israel. Middle East Journal, 11, 23-36.

Whitrow, G.J. 1972. Reflections on the history of the concept of time. In: J.T. Fraser, J.F.C. Haber & G.H. Mueller (eds.), The Study of Time. Berlin: Springer Verlag, pp. 1-11.

Wilson, M.G. 1979. The American Woman in Transition: The Urban Influence 1870-1920. Westport, CT: Greenwood Press.

Wilson, N.L. (1955). Space, time and individuals. Journal of Philosophy, 53, 589-98.

Wise, A. 1971. The impact of electronic communications on metropolitan form. Ekistics, 188, 22-31.

Women and Geography Study Group of the IBG, 1984. Geography and Gender. London: Hutchinson.

Zaretsky, E. 1976. Capitalism, the Family and Personal Life. New York: Harper Colophon Books.

Zelinsky, W., Monk, J., & Hanson, S. 1982. Women and geography: a review and perspectus. Progress in Human Geography, 6, 317-366.

Zentner, H. 1966. The social time-space relationship: a theoretical formulation. Sociological Inquiry, 36, 61-79.

Zerubavel, E. 1979. Private and public time: the temporal structure of social accessibility and professional commitments. Social Forces, 57, 38-58.

Zerubavel, E. 1981. Hidden Rhythms. Chicago: University of Chicago Press.

Index

The GeoJournal Library

Bruce Currey and Graeme Hugo (eds.), Famine as Geographical Phenomenon, 1984. ISBN 90–277–1762–1.

S. H. U. Bowie, F.R.S. and I. Thornton (eds.), Environmental Geochemistry and Health, 1985. ISBN 90–277–1879–2.

Leszek A. Kosinski and K. Maudood Elahi (eds.), Population Redistribution and Development in South Asia, 1985. ISBN 90–277–1938–1.

Yehuda Gradus (ed.), Desert Development, 1985. ISBN 90–277–2043–6.

Frank J. Calzonetti and Barry D. Solomon (eds.), Geographical Dimensions of Energy, 1985. ISBN 90–277–2061–4.

Jan Lundqvist, Ulrik Lohm and Malin Falkenmark (eds.), Strategies for River Basin Management, 1985. ISBN 90–277–2111–4.

Andrei Rogers and Frans J. Willekens (eds.), Migration and Settlement. A Multiregional Comparative Study, 1986. ISBN 90–277–2119–X.

Risto Laulajainen, Spatial Strategies in Retailing, 1987. ISBN 90–277–2595–0.

T. H. Lee, H. R. Linden, D. A. Dreyfus and T. Vasko (eds.), The Methane Age, 1988. ISBN 90–277–2745–7.

H. J. Walker (ed.), Artificial Structures and Shorelines, 1988. ISBN 90–277–2746–5.

Aharon Kellerman, Time, Space, and Society: Geographical Societal Perspectives, 1989. ISBN 0–7923–0123–4.